LATTICE THEORY
First Concepts and Distributive Lattices

GEORGE GRÄTZER
University of Manitoba

DOVER PUBLICATIONS, INC.
Mineola, New York

Bibliographical Note

This Dover edition, first published in 2009, is an unabridged republication of the work originally published in 1971 by W. H. Freeman and Company, San Francisco.

Library of Congress Cataloging-in-Publication Data

Gratzer, George A.
    Lattice theory : first concepts and distributive lattices / George Gratzer — Dover ed.
        p. cm.
    Originally published: San Francisco : W.H. Freeman, 1971.
    Includes bibliographical references and index.
    ISBN-13: 978-0-486-47173-0
    ISBN-10: 0-486-47173-X
    1. Lattice theory. 2. Lattices, Distributive. I. Title.

QA171.5.G73 2009
511.3'3—dc22

                                                                2008046925

Manufactured in the United States by LSC Communications
47173X03   2022
www.doverpublications.com

CONTENTS

2 DISTRIBUTIVE LATTICES

3 DISTRIBUTIVE LATTICES WITH PSEUDOCOMPLEMENTATION

PREFACE

In the first half of the nineteenth century, George Boole's attempt to formalize propositional logic led to the concept of Boolean algebras. While investigating the axiomatics of Boolean algebras at the end of the nineteenth century, Charles S. Peirce and Ernst Schröder found it useful to introduce the lattice concept. Independently, Richard Dedekind's research on ideals of algebraic numbers led to the same concept. In fact, Dedekind also introduced modularity, a weakened form of distributivity. Although some of the early results of these mathematicians and of Edward V. Huntington are very elegant and far from trivial, they did not attract the attention of the mathematical community.

It was Garrett Birkhoff's work in the mid-thirties that started the general development of lattice theory. In a brilliant series of papers he demonstrated the importance of lattice theory and showed that it provides a unifying framework for hitherto unrelated developments in many mathematical disciplines. Birkhoff himself, Valère Glivenko, Karl Menger, John von Neumann, Oystein Ore, and others had developed enough of this new field for Birkhoff to attempt to "sell" it to the general mathematical community, which he did with astonishing success in the first edition of his *Lattice Theory*. The further development of the subject matter can best be followed by comparing the first, second, and third editions of the book (G. Birkhoff [1940], [1948], and [1967]).

Distributive lattices have played many roles in the development of lattice theory. Historically, lattice theory started with (Boolean) distributive lattices; as a result, the theory of distributive lattices is the most extensive and most satisfying chapter in the history of lattice theory. Distributive lattices have provided the motivation for many results in general lattice theory. Many conditions on lattices and on elements and ideals of lattices are weakened forms of distributivity. Therefore, a thorough knowledge of distributive lattices is indispensable for work in lattice theory. Finally, in many applications the condition of distributivity is imposed on lattices arising in various areas of mathematics, especially algebra.

The realization of the special role of distributive lattices moved me to break with the traditional approach to lattice theory, which proceeds from partially ordered sets to general lattices, semimodular lattices, modular lattices, and finally distributive lattices. The goal of the present volume is to provide a detailed presentation of the theory of distributive lattices. General lattices will be discussed in the companion volume of the present one, which is now in preparation.

Chapter 1 includes a concise development of the basic concepts of lattice theory and a detailed development of free lattices. Diagrams are emphasized because I believe that an important part of learning lattice theory is the acquisition of skill in drawing diagrams. This chapter lays the foundation for the material included in both first and second volumes. Chapter 2, which develops the theory of distributive lattices, comprises the most substantial part of the present book. The last decade witnessed the birth of a new field in lattice theory: distributive lattices with pseudocomplementation. Chapter 3 discusses the basic results of this field and focuses on one of the major preoccupations of twentieth-century mathematicians: structure theory.

The exercises, which number more than 500, form an integral part of this book. Only exercises marked by an * could be left out by the reader without any loss in comprehension of the subject matter. The Bibliography contains only works that are referred to in the text. The sixty-seven original research problems, as well as the "Further Topics and References" included at the end of each chapter, should be of help to those who are interested in further reading and research in lattice theory.

Finally, the reader will note that the symbol ● is placed at the end of a proof; if a theorem or lemma contains more than one statement, the proof of a part is ended with ◗. The abbreviation "iff " stands for "if and only if." More difficult exercises are marked by *. "Theorem 10" refers to Theorem 10 of the same section, whereas "Theorem 15.10" refers to Theorem 10 of section 15. Similarly, "exercise 2.6" means exercise 6 of section 2. References to the Bibliography are given in the form "J. Jakubik [1957]," which refers to a paper (or book) by J. Jakubik published in 1957. Such references as "[1957a]" and "[1957b]" indicate that the Bibliography contains more than one work by the author published in that year. "R. S. Pierce [a]" refers to a paper by R. S. Pierce that had not appeared in print at the time the manuscript of this book was submitted for publication.

Winnipeg, Manitoba
September 1970   *George Grätzer*

ACKNOWLEDGMENTS

An individual takes the responsibility and gets (he hopes) the credit for writing a book, but in reality ventures of this kind often require the cooperation of a group of people.

The first version of this book was developed as a set of lecture notes for my course on lattice theory at the University of Manitoba in the academic year 1968–1969. The undergraduates and graduate students who took this course and many of my colleagues who attended helped by criticizing my lectures and by simplifying proofs. The lecture notes were also read by Dr. P. Burmeister, who offered many helpful remarks.

A rewritten version of the lecture notes was read by R. Balbes, M. I. Gould, K. M. Koh, H. Lakser, S. M. Lee, P. Penner, C. R. Platt, and R. Padmanabhan, who coordinated the work of this group. I rewrote a substantial part of the manuscript as a result of the changes they suggested. I am very thankful to the whole group, and especially to Dr. Platt and Dr. Padmanabhan, for their untiring work.

I would like to thank K. D. Magill, Jr., for the opportunity to conduct a course on lattice theory at the State University of New York at Buffalo in the summer of 1970, and the students of this class, especially J. H. Hoffman, for their valuable observations. Last but not least, the manuscript was read for the publisher by B. Jónsson; all his suggestions were gratefully accepted. The typing and retyping of the manuscript were done with unusual skill and patience by Mrs. M. McTavish. The rest of the secretarial work was capably handled by Mrs. N. Buckingham. The galley proofs were read by R. Antonius, J. A. Gerhard, K. M. Koh, W. A. Lampe, R. Quackenbush, I. Rival, and the group coordinator, R. Padmanabhan. E. Fried assisted me in compiling the Index.

Thanks are also due to the National Research Council of Canada for sponsoring much of the original research that has gone into this book and to Professor N. S. Mendelsohn, who relieved me of all teaching and administrative duties to allow me time to conduct research, to supervise the research of graduate students, and to prepare the manuscript of this book.

TABLE OF NOTATION

SETS

SYMBOL	EXPLANATION
$\in$	membership sign
$\subseteq$	inclusion
$\subset$	proper inclusion
$\cup, \bigcup$	union
$\cap, \bigcap$	intersection
$\varnothing$	empty set
$-$	set difference
$\times, \Pi$	Cartesian product
$|A|$	cardinality of the set A
$\varphi\psi, \varphi\cdot\psi$	composition of maps: $x(\varphi\psi) = (x\varphi)\psi$
φ^{-1}	inverse of a map
$A\varphi^{-1}$	$A\varphi^{-1} = \{x \mid x\varphi \in A\}$
$P(X)$	set of all subsets of X
ω	first infinite ordinal; $n < \omega$ means $n = 0, 1, 2, \ldots$
$\aleph_0$	first infinite cardinal
$\mathfrak{m}, \mathfrak{n}, \ldots$	lower-case German letters stand for cardinals

POSETS

SYMBOL	APPEARS ON PAGE(S)
$\leq$	1, 2
$\geq$	2
$\leq_B$	8
$\leq_\wedge$	9
$\leq_\vee$	9
$<$	8
$\prec$	12
$\succ$	12
$\parallel$	2
inf	3
sup	2
$\wedge$	3
$\vee$	2
$\cong$	19
$[a, b]$	20
$(a, b]$	20
$[a, b)$	20
(a, b)	20
γ	10
P^m	52
0	56
1	56

LATTICES

SPECIAL LATTICES AND ALGEBRAS

CLASSES OF LATTICES AND ALGEBRAS

CONGRUENCES

SUBSETS AND FAMILIES OF SUBSETS OF LATTICES

SYMBOL	APPEARS ON PAGE(S)
$[a,b]$	20
$(a, b]$	20
$[a, b)$	20
(a, b)	20
C^0	109
$D(L)$	162
F_a	163
$[H]$	20, 138
$(H]$	21
$[H)$	23
$[H]_R$	105
$[H]_B$	141
$H(P)$	72
I^*	107
$J(L)$	72
$\mathscr{P}(L)$	76, 119
$r(a)$	72, 75, 119
$r(I)$	119
$S(a)$	61
$S(I(L))$	107
$S(L)$	161
$X^\wedge$	133
$X^\vee$	133

OPERATIONS

$a \wedge b$	4
$a \vee b$	4
$a \vee b$	58
$x + y$	84, 92
a'	58
a^*	58
a^+	184

TOPOLOGY

$\bar{X}$	125
$\mathscr{S}(L)$	119

SYMBOL	APPEARS ON PAGE(S)
$\mathscr{S}^B$	124

ASSOCIATED ALGEBRAS

$A(L)$	64
$E(L)$	64
$E_{0,1}(L)$	64

PROPERTIES

(D)	50
(E)	41
(I)	50
(JID)	107
(L1)–(L4)	5
(L_n)	167
(MID)	107
(P1)–(P4)	1
$P(m, n)$	139
$P_{(0, 1)}(m, n)$	139
$P_B(m, n)$	139
(S1)	121
(S2)	121
(S3)	123
(S4)	187

MISCELLANEOUS

$\mathscr{C}(J)$	134
$\mathscr{C}_{red}(J)$	134
$f(k, n)$	64
$f_K(k, n)$	64
$\mathbf{H}$	175
$\mathbf{I}$	175
$\mathbf{P}^{(n)}$	32
$\mathbf{S}$	175
φ_θ	26

FIRST CONCEPTS

1. Two Definitions of Lattices

Whereas the arithmetical properties of the set of reals, R, can be expressed in terms of addition and multiplication, the order theoretic, and thus the topological, properties are expressed in terms of the ordering relation $\leq$. The basic properties of this relation are as follows.

For all $a,b,c \in R$ we have:

(P1)	$a \leq a$	(*reflexivity*)
(P2)	$a \leq b$ and $b \leq a$ imply that $a = b$	(*antisymmetry*)
(P3)	$a \leq b$ and $b \leq c$ imply that $a \leq c$	(*transitivity*)
(P4)	$a \leq b$ or $b \leq a$	(*linearity*)

There are many examples of binary relations sharing properties (P1)–(P4) with the ordering relation of reals, and there are even more enjoying (P1)–(P3). This fact, by itself, would not justify the introduction of a new concept. However, it has been shown that many basic concepts and results about the reals depend only on (P1)–(P3), and these can be profitably used whenever we have a relation satisfying (P1)–(P3). Relations satisfying (P1)–(P3) are called *partial ordering relations*, and sets equipped with such relations are called *partially ordered sets* or *posets*.

To make the definitions formal, let us start with two sets A, B and form the set $A \times B$ of all ordered pairs $\langle a, b \rangle$ with $a \in A$, $b \in B$. If $A = B$, we write A^2 for $A \times A$. Then a *binary relation* ρ on A can simply be defined as a subset of A^2. The elements a, b $(a, b \in A)$ are *in relation* with respect to ρ if $\langle a, b \rangle \in \rho$. For $\langle a, b \rangle \in \rho$, we will also write $a \rho b$, or $a \equiv b(\rho)$. Binary relations will be denoted by small Greek letters or by special symbols.

This formal definition can be compared with the intuitive one: A binary relation ρ on A is a "rule" that decides whether or not $a \rho b$ for any given $a, b \in A$. Of course, any such rule will determine the set $\{\langle a, b \rangle \mid a \rho b, a, b \in A\}$, and this set determines ρ, so we might as well regard ρ as being identical with this set.

A *partially ordered set* $\langle A; \rho \rangle$ consists of a nonvoid set A and a binary relation ρ on A, such that ρ satisfies properties (P1)–(P3). Note that these can be stated as follows: For all $a, b, c \in A$, $\langle a, a \rangle \in \rho$; $\langle a, b \rangle$, $\langle b, a \rangle \in \rho$ imply that $a = b$; $\langle a, b \rangle$, $\langle b, c \rangle \in \rho$ imply that $\langle a, c \rangle \in \rho$.

If ρ satisfies (P1)–(P3), ρ is a *partial ordering relation*, and it will usually be denoted by $\leq$. Also, $a \geq b$ will mean $b \leq a$. Sometimes we will say that A (rather than $\langle A; \leq \rangle$) is a poset, meaning that the partial ordering is understood. This is an ambiguous although widely accepted practice.

A poset $\langle A; \leq \rangle$ that also satisfies (P4) is called a *chain* (also called *fully ordered set, linearly ordered set*, and so on).

If $\langle A; \leq \rangle$ is a poset, $a, b \in A$, then a and b are *comparable* if $a \leq b$ or $b \leq a$. Otherwise, a and b are *incomparable*, in notation $a \parallel b$. A chain is, therefore, a poset in which there are no incomparable elements.

Next we define inf and sup in an arbitrary poset P (that is, $\langle P; \leq \rangle$) the same way as it is done for reals.

Let $H \subseteq P$, $a \in P$. Then a is an *upper bound* of H, if $h \leq a$ for all $h \in H$. An upper bound a of H is the *least upper bound* of H or *supremum* of H if, for any upper bound b of H, we have $a \leq b$. We shall write $a = \sup H$, or $a = \bigvee H$. (The notations $a = \text{l.u.b. } H$, $a = \sum H$ are also common in the literature.) This notation can be justified only if we show the uniqueness of the supremum. Indeed, if a_0 and a_1 are both suprema of H, then $a_0 \leq a_1$, since a_1 is an upper bound, and a_0 is a supremum. Similarly, $a_1 \leq a_0$; thus $a_0 = a_1$ by (P2).

The concepts of *lower bound* and *greatest lower bound* or *infimum* are similarly defined; the latter is denoted by inf H or $\bigwedge H$. (The notations g.l.b. H, $\prod H$ are also used in the literature.) The uniqueness is proved as in the preceding paragraph.

The adverb "similarly" at the end of that paragraph can be given a very concrete meaning. Let $\langle P; \leq \rangle$ be a poset. The notation $a \geq b$ (meaning $b \leq a$) can also be regarded as a definition of a binary relation on P. This binary relation $\geq$ satisfies (P1)–(P3); as an example, let us check (P2). If $a \geq b$ and $b \geq a$, then by the definition of $\geq$ we have $b \leq a$ and $a \leq b$; using (P2) for $\leq$ we conclude that $a = b$. (P1) and (P3) are equally trivial. Thus $\langle P; \geq \rangle$ is also a poset, called the *dual* of $\langle P; \leq \rangle$. Now, if Φ is a "statement" about posets, and if in Φ we replace all occurrences of $\leq$ by $\geq$, we get the *dual* of Φ.

DUALITY PRINCIPLE.   *If a statement Φ is true in all posets, then its dual is also true in all posets.*

This is true simply because Φ holds for $\langle P; \leq \rangle$ iff the dual of Φ holds for $\langle P; \geq \rangle$, which is also a poset.

As an example take for Φ the statement: "If sup H exists it is unique." We get as its dual: "If inf H exists it is unique."

It is hard to imagine that anything as trivial as the Duality Principle could yield anything profound, and it does not; but it can save a lot of work.

A poset $\langle L; \leq \rangle$ is a *lattice* if sup $\{a, b\}$ and inf $\{a, b\}$ exist for all $a,b \in L$.

In other words, lattice theory singles out a special type of poset for detailed investigation. To make such a definition worthwhile, it must be shown that this class of posets is a very useful class, that there are many such posets in various branches of mathematics (analysis, topology, logic, algebra, geometry), and that a general study of these posets will lead to a better understanding of the behavior of the examples. This was done in the first edition of G. Birkhoff's *Lattice Theory* [1940]. As we go along we shall see many examples, most of them in the exercises. For a general survey of lattices in mathematics, see G. Birkhoff [1967], and H. H. Crapo and G. C. Rota [1970].

We will take the usefulness of lattice theory for granted (this is a less touchy subject now than it formerly was) and hope that the reader will like it for its intrinsic beauty.

To make the definition of a lattice less arbitrary, we note that an equivalent definition is the following:

A poset $\langle L; \leq \rangle$ *is a lattice if* sup H *and* inf H *exist for any finite nonvoid subset H of L.*

PROOF. It is enough if we prove that the first definition implies the second. So let $\langle L; \leq \rangle$ satisfy the first definition and let $H \subseteq L$ be nonvoid and finite. If $H = \{a\}$, then sup $H =$ inf $H = a$ follows from the reflexivity of $\leq$ and the definitions of sup and inf. Let $H = \{a, b, c\}$. To show that sup H exists, set $d =$ sup $\{a, b\}$, $e =$ sup $\{d, c\}$; we claim that $e =$ sup H. First of all, $a \leq d, b \leq d$, and $d \leq e, c \leq e$; therefore (by transitivity) $x \leq e$, for all $x \in H$. Secondly, if f is an upper bound of H, then $a \leq f, b \leq f$, and thus $d \leq f$; also $c \leq f$, so that $c,d \leq f$; therefore $e \leq f$, since $e =$ sup $\{d,c\}$. Thus e is the supremum of H.

If $H = \{a_0, \ldots, a_{n-1}\}, n \geq 1$, then

$$\text{sup} \{ \ldots \text{sup} \{\text{sup} \{a_0, a_1\}, a_2\} \ldots, a_{n-1}\}$$

is the supremum of H, by an inductive argument whose steps are analogous to those in the preceding paragraph.

By duality (in other words, by applying the Duality Principle), we conclude that inf H exists. ●

With regard to $H = \varnothing$ (the void set), we will point out in the exercises that inf $\varnothing$ and sup $\varnothing$ need not exist in a lattice.

This simple proof can be varied to yield a large number of equally trivial statements about lattices and partially ordered sets in general. Some of these will be stated as exercises and used later. To make the use of the Duality Principle legitimate for lattices, note:

If $\langle P; \leq \rangle$ is a lattice, so is its dual $\langle P; \geq \rangle$.

Thus the Duality Principle applies to lattices.

We will use the notations

$$a \wedge b = \text{inf} \{a, b\}$$

and

$$a \vee b = \text{sup} \{a, b\}$$

and call $\wedge$ the *meet* and $\vee$ the *join*. In lattices, they are both *binary operations*, which means that they can be applied to a pair of elements

a, b of L to yield again an element of L. Thus $\wedge$ is a map of L^2 into L and so is $\vee$, a remark that might fail to be very illuminating at this point. The previous proof yields that

$$(\cdots((a_0 \vee a_1) \vee a_2)\cdots) \vee a_{n-1} = \sup\{a_0, \ldots, a_{n-1}\}$$

and there is a similar formula for inf. Now observe that the right-hand side does not depend on the way the elements a_i are listed. Thus $\wedge$ and $\vee$ are idempotent, commutative, and associative—that is,

(L1) $a \wedge a = a, a \vee a = a$ (*idempotency*)

(L2) $a \wedge b = b \wedge a, a \vee b = b \vee a$ (*commutativity*)

(L3) $(a \wedge b) \wedge c = a \wedge (b \wedge c)$,

        $(a \vee b) \vee c = a \vee (b \vee c)$ (*associativity*)

As always in algebra, associativity makes it possible to write

$$a_0 \wedge a_1 \wedge \cdots \wedge a_{n-1}$$

without using parentheses (and the same for $\vee$).

There is another pair of rules that connect $\wedge$ and $\vee$. To derive them, note that if $a \leq b$, then inf $\{a, b\} = a$; that is, $a \wedge b = a$, and conversely. Thus

$$a \leq b \text{ iff } a \wedge b = a.$$

By duality (and by interchanging a and b) we have

$$a \leq b \text{ iff } a \vee b = b.$$

Applying the "only if" part of the first rule to a and $a \vee b$, and that of the second rule to $a \wedge b$ and a, we get

(L4) $a \wedge (a \vee b) = a, a \vee (a \wedge b) = a$ (*absorption identities*)

Now we are faced with the crucial question: Do we know enough about $\wedge$ and $\vee$ so that lattices can be characterized purely in terms of properties of $\wedge$ and $\vee$?

Two comments are in order. It is obvious that $\leq$ can be characterized by $\wedge$ and $\vee$ (in fact, by either of them); therefore, obtaining such a characterization is only a matter of persistence. More importantly, why should we try to get such a characterization? To rephrase the question, why should we want to characterize $\langle L; \leq \rangle$ as $\langle L; \wedge, \vee \rangle$, which is an algebra—that is, a set equipped with operations (in this case, two binary operations)? Note that $\leq$ is simply a subset of L^2, whereas $\wedge$ and $\vee$ are maps from L^2 into L. The answer is simple: We want such a characterization because if we can treat lattices as algebras, then all the concepts and methods of universal algebra will become applicable. The usefulness of treating lattices as algebras will soon become clear.

An algebra $\langle L; \wedge, \vee \rangle$ is called a *lattice* if: L is a nonvoid set, $\wedge$ and $\vee$ are binary operations on L, both $\wedge$ and $\vee$ are idempotent, commutative, and associative, and they satisfy the two absorption identities. The following theorem states that a lattice as an algebra and a lattice as a poset are "equivalent" concepts. (The word "equivalent" will not be defined.)

THEOREM 1.

(i) *Let the poset* $\mathfrak{L} = \langle L; \leq \rangle$ *be a lattice. Set* $a \wedge b = inf\{a, b\}$, $a \vee b = sup\{a, b\}$. *Then the algebra* $\mathfrak{L}^a = \langle L; \wedge, \vee \rangle$ *is a lattice.*

(ii) *Let the algebra* $\mathfrak{L} = \langle L; \wedge, \vee \rangle$ *be a lattice. Set* $a \leq b$ *iff* $a \wedge b = a$. *Then* $\mathfrak{L}^p = \langle L; \leq \rangle$ *is a poset, and the poset* $\mathfrak{L}^p$ *is a lattice.*

(iii) *Let the poset* $\mathfrak{L} = \langle L; \leq \rangle$ *be a lattice. Then* $(\mathfrak{L}^a)^p = \mathfrak{L}$.

(iv) *Let the algebra* $\mathfrak{L} = \langle L; \wedge, \vee \rangle$ *be a lattice. Then* $(\mathfrak{L}^p)^a = \mathfrak{L}$.

REMARK. (i) and (ii) describe how we pass from poset to algebra and back, whereas (iii) and (iv) state that going there and back takes us back to where we started.

PROOF.

(i) This has already been proved. ▶

(ii) We set $a \leq b$ to mean $a \wedge b = a$. Now $\leq$ is reflexive since $\wedge$ is idempotent; $\leq$ is antisymmetric since $a \leq b$, $b \leq a$ mean that $a \wedge b = a, b \wedge a = b$, which, by the commutativity of $\wedge$, imply that $a = a \wedge b = b \wedge a = b$; $\leq$ is transitive, since if $a \leq b$, $b \leq c$, then $a = a \wedge b, b = b \wedge c$, and so

$$a = a \wedge b = a \wedge (b \wedge c) \ (\wedge \text{ is associative})$$

$$= (a \wedge b) \wedge c = a \wedge c,$$

that is, $a \le c$. Thus $\langle L; \le \rangle$ is a poset. To prove that $\langle L; \le \rangle$ is a lattice we will verify that $a \wedge b = \inf\{a, b\}$ and $a \vee b = \sup\{a, b\}$. (This is not a definition.) Indeed, $a \wedge b \le a$, since

$$(a \wedge b) \wedge a = a \wedge (b \wedge a) = a \wedge (a \wedge b)$$

$$= (a \wedge a) \wedge b = a \wedge b,$$

using associativity, commutativity, and idempotency of $\wedge$; similarly, $a \wedge b \le b$. Now if $c \le a$, $c \le b$, that is, $c \wedge a = c$, $c \wedge b = c$, then $c \wedge (a \wedge b) = (c \wedge a) \wedge b = c \wedge b = c$; thus $a \wedge b = \inf\{a, b\}$. Finally, $a \le a \vee b$, $b \le a \vee b$, because $a = a \wedge (a \vee b)$, $b = b \wedge (a \vee b)$ by the first absorption identity; if $a \le c$, $b \le c$, that is, $a = a \wedge c$, $b = b \wedge c$, then $a \vee c = (a \wedge c) \vee c = c$, and $b \vee c = c$ by the second absorption identity. Thus

$$(a \vee b) \wedge c = (a \vee b) \wedge (a \vee c) = (a \vee b) \wedge [a \vee (b \vee c)]$$

$$= (a \vee b) \wedge [(a \vee b) \vee c] = a \vee b,$$

that is, $a \vee b \le c$, completing the proof of $a \vee b = \sup\{a, b\}$. ▌

(iii) It is enough to observe that the partial ordering of $\mathfrak{L}$ and $(\mathfrak{L}^a)^p$ are identical to get (iii).

(iv) The proof of (iv) is similar to the proof of (iii). ●

The proof of Theorem 1, and even the statement of Theorem 1, are subject to criticism. To begin with, in the definition of a lattice as an algebra, idempotency is redundant. The last step of the proof of (ii) can be made neater by first proving that

$$a = a \wedge b \text{ iff } b = a \vee b.$$

Theorem 1 should be preceded by a similar theorem for "semilattices." All these questions will be dealt with in the exercises that follow this section.

Finally, note that for lattices as algebras, the Duality Principle takes on the following very simple form.

Let Φ be a statement about lattices expressed in terms of $\wedge$ and $\vee$. The dual of Φ is the statement we get from Φ by interchanging $\wedge$ and $\vee$. If Φ is true for all lattices, then the dual of Φ is also true for all lattices.

To prove this we have only to observe that if $\mathfrak{L} = \langle L; \wedge, \vee \rangle$, then the dual of $\mathfrak{L}^p$ is $(\langle L; \vee, \wedge \rangle)^p$.

Exercises

Posets

1. Define $x < y$ to mean $x \leq y$ and $x \neq y$. Prove that, in a partially ordered set, $x < x$ for no x, and that $x < y$, $y < z$ imply that $x < z$.
2. Let the binary relation $<$ satisfy the conditions of exercise 1. Define $x \leq y$ to mean $x < y$ or $x = y$. Then show that $\leq$ is a partial ordering.
3. Prove the following extension of antisymmetry: If $x_0 \leq x_1 \leq \cdots \leq x_{n-1} \leq x_0$, then $x_0 = x_1 = \cdots = x_{n-1}$.
4. Enumerate all partial orderings on a five-element set.
5. Let $\leq$ be a partial ordering on A and let B be a subset of A. Set $a \leq_B b$ for $a,b \in B$ if $a \leq b$. Prove that $\leq_B$ is a partial ordering on B.
6. Let A be a set and let P be the set of all partial orderings on A. For $\rho, \sigma \in P$ set $\rho \leq \sigma$ if $a \rho b$ implies that $a \sigma b$ $(a,b \in A)$. Prove that $\langle P; \leq \rangle$ is a poset.
7. Let $\varnothing$ denote the void set. If inf $\varnothing = a$, then $x \leq a$ for all elements x of the poset. Find an example of a poset in which inf $\varnothing$ does not exist.
8. Formulate and prove exercise 7 for sup $\varnothing$.

Lattices as Posets

9. Prove that the following are examples of posets:
   (i) Let $A = P(X)$, the set of all subsets of a set X; let $X_0 \leq X_1$ mean $X_0 \subseteq X_1$ (set inclusion).
   (ii) Let A be the set of all real valued functions defined on X; for $f,g \in A$ set $f \leq g$ iff $f(x) \leq g(x)$ for all $x \in X$.
   (iii) Let A be the set of all continuous concave real valued functions defined on the real interval; define $f \leq g$ as in (ii).
   (iv) Let A be the set of all open sets of a topological space; define $\leq$ as in (i).
   (v) Let A be the set of all human beings; $a < b$ means that a is a descendant of b.
10. Which of the examples in exercise 9 are lattices? For those that are lattices, compute $a \wedge b$, $a \vee b$.
11. Show that every chain is a lattice.
12. Let A be the set of all subgroups (normal subgroups) of a group G; for $X,Y \in A$ set $X \leq Y$ to mean $X \subseteq Y$. Prove that $\langle A; \leq \rangle$ is a lattice; compute $X \wedge Y$ and $X \vee Y$.
13. Let $\langle P; \leq \rangle$ be a poset in which inf H exists for *all* $H \subseteq P$. Show that $\langle P; \leq \rangle$ is a lattice. (Hint: For $a,b \in P$ let H be the set of all upper bounds of $\{a, b\}$. Prove that sup $\{a, b\}$ = inf H.) Relate this to exercise 12.

Semilattices as Posets

14. *A poset is a* join-semilattice (*dually,* meet-semilattice) *if* sup $\{a,b\}$ (*dually,* inf $\{a, b\}$) *exists for any two elements.*

Prove that the dual of a join-semilattice is a meet-semilattice, and conversely.

15. Let A be the set of finitely generated subgroups of a group G, partially ordered under set inclusion (as in exercise 12). Prove that $\langle A; \subseteq \rangle$ is a join-semilattice, but not necessarily a lattice.

16. Let C be the set of all continuous strictly convex real valued functions defined on the real interval $[0, 1]$. For $f, g \in C$ set $f \le g$ iff $f(x) \le g(x)$ for all $x \in [0, 1]$. Prove that $\langle C; \le \rangle$ is a meet-semilattice, but not a join-semilattice.

17. Show that the poset $\langle P; \le \rangle$ is a lattice iff it is a join- and meet-semilattice.

Semilattices as Algebras

18. Let $\langle A; \circ \rangle$ be an algebra with one binary operation $\circ$. The algebra $\langle A; \circ \rangle$ is a semilattice if $\circ$ is idempotent, commutative, and associative. Let $\langle P; \le \rangle$ be a join-semilattice. Show that the algebra $\langle A; \vee \rangle$ is a semilattice when $a \vee b = \sup \{a, b\}$. State the analogous result for meet-semilattices.

19. Let the algebra $\langle A; \circ \rangle$ be a semilattice. Define the binary relations $\le_\wedge$, $\le_\vee$ on A as follows: $a \le_\wedge b$ iff $a = a \circ b$; $a \le_\vee b$ iff $b = a \circ b$. Prove that $\langle A; \le_\wedge \rangle$ is a poset, as a poset it is a meet-semilattice, and $a \wedge b = a \circ b$; that $\langle A; \le_\vee \rangle$ is a poset, as a poset it is a join-semilattice, and $a \vee b = a \circ b$; and that the dual of $\langle A; \le_\wedge \rangle$ is $\langle A; \le_\vee \rangle$.

Theorem 1 for Semilattices

20. Prove the following statements:
    (i) Let the poset $\mathfrak{A} = \langle A; \le \rangle$ be a join-semilattice. Set $a \vee b = \sup \{a, b\}$. Then the algebra $\mathfrak{A}^a = \langle A; \vee \rangle$ is a semilattice.
    (ii) Let the algebra $\mathfrak{A} = \langle A; \circ \rangle$ be a semilattice. Set $a \le b$ iff $a \circ b = b$. Then $\mathfrak{A}^p = \langle A; \le \rangle$ is a poset, and the poset $\mathfrak{A}^p$ is a join-semilattice.
    (iii) Let the poset $\mathfrak{A} = \langle A; \le \rangle$ be a join-semilattice. Then $(\mathfrak{A}^a)^p = \mathfrak{A}$.
    (iv) Let the algebra $\mathfrak{A} = \langle A; \circ \rangle$ be a semilattice. Then $(\mathfrak{A}^p)^a = \mathfrak{A}$.

21. Formulate and prove Theorem 1 for the meet-semilattices.

Lattices as Algebras

22. Prove that the absorption identities imply the idempotency of $\wedge$ and $\vee$. (Hint: simplify $a \vee [a \wedge (a \vee a)]$ in two ways to yield $a = a \vee a$.)

23. Let the algebra $\langle A; \wedge, \vee \rangle$ be a lattice. Define $a \le_\wedge b$ iff $a \wedge b = a$; $a \le_\vee b$ iff $a \vee b = b$. Prove that $a \le_\wedge b$ iff $a \le_\vee b$.

24. Prove that the algebra $\langle A; \wedge, \vee \rangle$ is a lattice iff $\langle A; \wedge \rangle$ and $\langle A; \vee \rangle$ are semilattices and $a = a \wedge b$ is equivalent to $b = a \vee b$. Verify that if $\langle A; \wedge, \vee_1 \rangle$ and $\langle A; \wedge, \vee_2 \rangle$ are both lattices, then $\vee_1$ is the same as $\vee_2$.

25. Is there a result for semilattices that is analogous to that of exercise 22? In other words, are the three conditions for semilattices independent (that is, none follows from the others)?

*26. Prove that an algebra $\langle A; \wedge, \vee \rangle$ is a lattice iff it satisfies

$$a = (b \wedge a) \vee a$$

and

$$\{[(a \wedge b) \wedge c] \vee d\} \vee e = \{[(b \wedge c) \wedge a] \vee e\} \vee [(f \vee d) \wedge d]$$

(see J. A. Kalman [1968]. The first definition of lattices by two identities was found by R. Padmanabhan in 1967, *Notices Amer. Math. Soc.* 14, No. 67T-468, and published in full in R. Padmanabhan [1969]. J. A. Kalman's two identities are slightly improved versions of those of R. Padmanabhan. R. N. McKenzie [a] proved the existence of a single identity characterizing lattices.)

Miscellany

27. Let ρ be a binary relation on the set A. The *transitive extension of* ρ is a binary relation β defined by the following rule: $a \beta b$ iff there exists a sequence $a = x_0, x_1, \ldots, x_n = b$ with $x_i \rho x_{i+1}$ for $i = 0, \ldots, n - 1$. Show that for a reflexive relation ρ, the transitive extension β is a partial ordering relation iff β is antisymmetric. Express this condition in terms of ρ.

28. Let ρ be a reflexive and transitive binary relation (*quasi-ordering relation*) on the nonvoid set A. Call $B \subseteq A$ a *block* if $B \neq \varnothing$, $a \rho b$, and $b \rho a$ for any $a,b \in B$, and for $a \in B$, $b \in A$, $a \rho b$, and $b \rho a$ imply that $b \in B$. Let P be the set of all blocks and set $B_1 \leq B_2$ $(B_1, B_2 \in P)$ if $b_1 \rho b_2$ for some (thus for all) $b_1 \in B_1$, $b_2 \in B_2$. Prove that $\langle P; \leq \rangle$ is a poset.

29. Let A be a set of sets. Let $a \rho b$ mean that there is a one-to-one map from a into b. What is $\langle P; \leq \rangle$? (Notation is that of exercise 28.)

30. Let the binary operation $\circ$ on the set A be associative. Give a rigorous proof of the statement that any bracketing of $a_0 \circ \cdots \circ a_{n-1}$ will yield the same element.

31. Suppose that in a poset $b \vee c$, $a \vee (b \vee c)$, and $a \vee b$ exist. Prove that $(a \vee b) \vee c$ exists and that $a \vee (b \vee c) = (a \vee b) \vee c$.

32. Prove that if $a \wedge b$ exists in a poset, so does $a \vee (a \wedge b)$.

33. Let H and K be subsets of a poset. Suppose that sup H, sup K, and sup $(H \cup K)$ exist. ($H \cup K$ is the set union of H and K.) Under these conditions verify that $(\sup H) \vee (\sup K)$ exists and equals sup $(H \cup K)$.

34. In a poset P define the *comparability relation* γ: For $a,b \in P$, $a \gamma b$ if $a \leq b$ or $b \leq a$. Take a sequence $a_1, \ldots, a_k$ of elements of P satisfying the following conditions:

    (i) $a_i \neq a_{i+1}$, $a_i \gamma a_{i+1}$ for $i = 1, \ldots, k - 1$; $a_k \neq a_1$, $a_k \gamma a_1$.

    (ii) For no elements $a,b \in P$, and $i,j < k$, $i \neq j$ is $a = a_i = a_j$, $b = a_{i+1} = a_{j+1}$, or $a = a_i = a_k$, $b = a_{i+1} = a_1$.

    (iii) For no $1 \leq i \leq k - 2$ is $a_i \gamma a_{i+2}$, and neither $a_{k-1} \gamma a_1$ nor $a_k \gamma a_2$ holds.

Prove that k is even.

*35.   Prove that a binary relation γ on a set A is the comparability relation of some poset $\langle A; \leq \rangle$ iff γ satisfies the condition of exercise 34 (A. Ghouila-Houri [1962]; see also P. C. Gilmore and A. J. Hoffman [1964] and M. Aigner [1969]).

2. How to Describe Lattices

To illustrate results and to refute conjectures we shall have to describe a large number of examples of lattices. This can be done by basing the examples on known mathematical structures, as illustrated in the exercises of Section 1. In this section we list a few other methods.

A finite lattice can always be described by a *meet table* and a *join table*. For example: let $L = \{0, a, b, 1\}$.

$\wedge$	0	a	b	1		$\vee$	0	a	b	1
0	0	0	0	0		0	0	a	b	1
a	0	a	0	a		a	a	a	1	1
b	0	0	b	b		b	b	1	b	1
1	0	a	b	1		1	1	1	1	1

We see that most of the information provided by the tables is redundant. Since both operations are commutative, the tables are symmetric with respect to the diagonal. Furthermore, $x \wedge x = x$, $x \vee x = x$; thus the diagonals themselves do not provide new information. Therefore, the two tables can be condensed into one.

$\wedge$ $\vee$	0	a	b	1
0		0	0	0
a	a		0	a
b	b	1		b
1	1	1	1	

It should be emphasized again that the part above the diagonal determines the part below the diagonal since either determines the partial ordering. To show that this table defines a lattice, we have only to check the associative identities and the absorption identities.

An alternative way is to describe the partial ordering, that is, all pairs

Figure 2.1

$\langle x, y \rangle$ with $x \leq y$. In the preceding example we get $\leq = \{\langle 0, 0 \rangle, \langle 0, a \rangle,$ $\langle 0, b \rangle, \langle 0, 1 \rangle, \langle a, a \rangle, \langle a, 1 \rangle, \langle b, b \rangle, \langle b, 1 \rangle, \langle 1, 1 \rangle \}$.

Obviously, all pairs of the form $\langle x, x \rangle$ can be omitted from the list, since we know that $x \leq x$. Also, if $x \leq y$ and $y \leq z$, then $x \leq z$. For instance, when we know that $0 \leq a$ and $a \leq 1$, we do not have to be told that $0 \leq 1$. To make this idea more precise, let us say that in the poset $\langle P; \leq \rangle$, a covers b (b is *covered by* a) (in notation, $a \succ b$ ($b \prec a$)) if $a > b$ and for *no* x, $a > x > b$. The covering relation of the preceding example is simply:

$$\prec = \{\langle 0, a \rangle, \langle 0, b \rangle, \langle a, 1 \rangle, \langle b, 1 \rangle\}.$$

Does the covering relation determine the partial ordering? The following lemma shows that in the finite case it does.

LEMMA 1. *Let $\langle P; \leq \rangle$ be a finite poset. Then $a \leq b$ iff $a = b$ or there exists a finite sequence of elements $x_0, \ldots, x_{n-1}$ such that $x_0 = a, x_{n-1} = b$, and $x_i \prec x_{i+1}$, for $0 \leq i < n - 1$.*

PROOF. If there is such a finite sequence, then $a = x_0 \leq x_1 \leq \cdots \leq x_{n-1} = b$, and a trivial induction on n yields $a \leq b$. Thus it suffices to prove that if $a < b$, then there is such a sequence. Fix $a, b \in P$, $a < b$, and take all subsets H of P such that H is a chain (under the partial ordering induced by the partial ordering of P), a is the smallest element of H, and b is the largest element of H. There are such subsets: $\{a, b\}$, for example. Choose such an H with the largest possible number of elements, say, with m elements. (P is finite.) Then $H = \{x_0, \ldots, x_{m-1}\}$, and we can assume that $x_0 < x_1 < \cdots < x_{m-1}$. We claim that in $\langle P; \leq \rangle$ we have $a = x_0 \prec x_1 \prec \cdots \prec x_{m-1}$. Indeed, $x_i < x_{i+1}$ by assumption. Thus if $x_i \prec x_{i+1}$ does not hold, then $x_i < x < x_{i+1}$ for some $x \in P$, and $H \cup \{x\}$ will be a chain of $m + 1$ elements between a and b, contrary to the maximality of the number of elements of H. ●

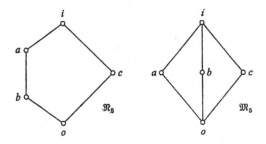

Figure 2.2

The *diagram* of a poset $\langle P; \leq \rangle$ represents the elements with small circles ∘; the circles representing two elements x, y are connected by a straight line if one covers the other; if x covers y, then the circle representing x is higher up than the circle representing y. The diagram of the lattice discussed previously is shown in Figure 2.1. Sometimes the "diagram" of an infinite poset is drawn. Such diagrams are always accompanied by explanations in the text. Note that in a diagram the intersection of two lines need not indicate an element. A diagram is *planar* if no two lines intersect. A diagram is *optimal* if the number of pairs of intersecting lines is minimal. Figures 2.1, 2.2, and 2.3 are planar diagrams; Figure 2.4 is an optimal but not planar diagram; Figure 2.6 is not optimal. As a rule, optimal diagrams are the most practical to use.

The methods we will use will be combinations of previous ones. The lattice $\mathfrak{N}_5$ of Figure 2.2 has five elements: o, a, b, c, i, and $b < a, c \vee b = i$, $a \wedge c = o$. This description is complete—that is, all the relations follow from the ones given. $\mathfrak{M}_5$ has five elements: o, a, b, c, i and $a \wedge b = a \wedge c = b \wedge c = o, a \vee b = a \vee c = b \vee c = i$.

In contrast, we can start with some elements (say, a, b, c), with some relations (say, $b < a$), and ask for the "most general" lattice that can be formed, *without* specifying the elements to be used. (The exact meaning of "most general" will be given in Section 5.) In this case we continue to form meets and joins until we get a lattice. A meet (or join) formed is identified with an element that we already have only if this identification is forced by the lattice axioms or by the given relations. The lattice we get from a, b, c, satisfying $b < a$ is shown in Figure 2.3.

To illustrate these ideas we give a part of the computation that goes into the construction of the most general lattice L generated by a, b, c with

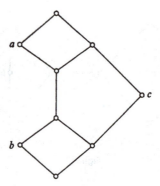

Figure 2.3

$b < a$. We start by constructing joins and meets $a \vee c$, $b \vee c$, $a \wedge c$, $b \wedge c$; note that $a \vee b \vee c = (a \vee b) \vee c = a \vee c$ since $a \vee b = a$; similarly, $a \wedge b \wedge c = b \wedge c$. Next we have to show that the seven elements $(a, b, c, a \vee c, b \vee c, a \wedge c, b \wedge c)$ that we already have are all distinct. Remember that two were equal if such equality would follow from the relation $(b < a)$ and the lattice axioms. Therefore, to show a pair of them distinct, it is enough to find a lattice K with $a, b, c \in K$, $a < b$, where that pair of elements is distinct. For instance, to show $a \neq a \vee c$, take the lattice $\{0, 1, 2\}$ with $0 < 1 < 2$, and $b = 0$, $a = 1$, $c = 2$. The next step is to form further joins and meets: $b \vee (a \wedge c), a \wedge (b \vee c)$. It is easy to see that all the other joins and meets are equal to a given one—for example, $b \wedge (a \wedge c) = b \wedge c$, $a \vee (a \wedge c) = a$. Now we claim that the nine elements $(a, b, c, a \vee c, b \vee c, a \wedge c, b \wedge c, b \vee (a \wedge c), a \wedge (b \vee c))$ form a lattice. We have to prove that by joins and meets we cannot get anything new. The thirty-six joins and thirty-six meets we have to check are all trivial. For instance, $a \vee [b \vee (a \wedge c)] = a \vee b \vee (a \wedge c) = a$, by the absorption law, since $a \vee b = a$; also $c \wedge [a \wedge (b \vee c)] = (c \wedge a) \wedge (b \vee c) = a \wedge c$, since $c \wedge a \leq b \vee c$.

Exercises

1. Give the join and meet table of the lattice in exercise 1.9(a) for one-, two-, and three-element sets X.

2. Give the set $\leq$ for the lattices in exercise 1. Which is simpler: the meet and join table or $\leq$?

3. Describe a practical method of checking associativity in a join (meet) table.

4. Relate Lemma 1 to exercise 1.27.

5. Let $\langle P; \leq \rangle$ be a poset, $a,b \in P$, $a < b$, and let C denote the set of all chains H in P with smallest element a, largest element b. Let $H_0 \leq H_1$ mean $H_0 \subseteq H_1$ for $H_0, H_1 \in C$. Show that $\langle C; \leq \rangle$ is a poset with smallest element $\{a, b\}$.

6. The poset $\langle Q; \leq \rangle$ is said to satisfy the *Ascending Chain Condition* if any increasing chain terminates, that is, if $x_i \in Q$, $i = 0, 1, 2, \ldots$, and $x_0 \leq x_1 \leq \cdots \leq x_n \leq \cdots$, then for some m we have $x_m = x_{m+1} = \cdots$. The element x of Q is *maximal* if $x \leq y \, (y \in Q)$ implies that $x = y$. Show that the Ascending Chain Condition implies the existence of maximal elements and that, in fact, every element is included in a maximal element.

7. Dualize exercise 6. (The dual of maximal is *minimal* and the dual of ascending is *descending*.)

8. If $\langle Q; \leq \rangle$ is a lattice and x is a maximal element, then $y \leq x$ for all $y \in Q$. Show that this statement is not, in general, true in a poset.

9. Give examples of posets without maximal elements and of posets with maximal elements in which not every element is included in a maximal element.

10. Let $\langle P; \leq \rangle$ be a poset with the property that for every $a,b \in P$, $a < b$, any chain in P with smallest element a and largest element b is finite. Show that the poset $\langle C; \leq \rangle$ formed from $\langle P; \leq \rangle$ in exercise 5 satisfies the Ascending Chain Condition (see exercise 7).

11. Extend Lemma 1 to posets satisfying the condition of exercise 10 (combine exercises 6 and 10).

12. Could the result of exercise 11 be proved using the reasoning of Lemma 1?

13. Is the result of exercise 11 the best possible? (No)

14. Describe a method of finding the meet and join table of a lattice given by a diagram.

15. Are the posets of Figures 2.4 and 2.5 lattices?

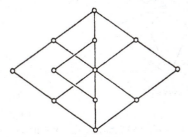
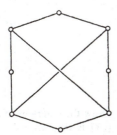

Figure 2.4   Figure 2.5

16.   Show that Figures 2.6 and 2.7 represent the same lattice.

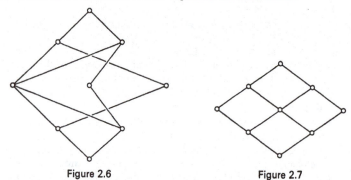

Figure 2.6   Figure 2.7

17.   Simplify Figure 2.8.

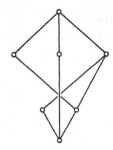

Figure 2.8

18.   Simplify Figure 2.9. What is the number of pairs of intersecting lines in an optimal diagram? (Zero)

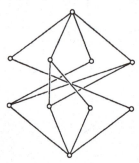

Figure 2.9

19.   Draw the diagrams of $P(X)$ (exercise 1.9(a)) for $|X| = 3$ and $|X| = 4$. ($|X|$ is the cardinality of X.) Does $P(X)$ have a planar diagram for $|X| = 3$?

20. Draw the diagrams of the lattice of binary relations on X (partially ordered by set inclusion) for $|X| \leq 4$.

21. Describe the most general lattice generated by a, b, c such that $a > b$, $a \vee c = b \vee c, a \wedge c = b \wedge c$.

*22. Describe the most general lattice generated by a, b, c, d, such that $a > b > c$. (The diagram is given in Ju. I. Sorkin [1952]; the lattice has twenty elements.)

*23. Show that the most general lattice generated by a, b, c, d, such that $a > b, c > d$, is infinite. (This lattice is described in H. L. Rolf [1958].)

24. Let N be the set of positive integers, $L = \{\langle n, i \rangle \mid n \in N, i = 0, 1\}$. Set $\langle n, i \rangle \leq \langle m, j \rangle$ if $n \leq m$ and $i \leq j$. Show that L is a lattice and draw the "generalized diagram" of L.

25. Draw the diagrams of all lattices with, at most, six elements.

To dispel a false impression that may have been created by Sections 1 and 2, namely, that the proof that a poset is a lattice is always trivial, we present exercises 26–36 showing that the poset T_n defined in exercise 34 is a lattice. This is a result of D. Tamari [1951], first published in H. Friedman and D. Tamari [1967]. The present proof is based on S. Huang and D. Tamari [a].

26. Let T_n denote the set of all possible *binary bracketings* of $x_0 x_1 \cdots x_n$; for instance, $T_0 = \{x_0\}$, $T_1 = \{(x_0 x_1)\}$, $T_2 = \{((x_0 x_1)x_2), (x_0(x_1 x_2))\}$, $T_3 = \{(x_0(x_1(x_2 x_3))), (x_0((x_1 x_2)x_3)), ((x_0 x_1)(x_2 x_3)), ((x_0(x_1 x_2))x_3), (((x_0 x_1)x_2)x_3)\}$. Give a formal (inductive) definition of T_n.

27. Replacing consistently all occurrences of $(A \cdot B)$ by $A(B)$ in a binary bracketing, we get the *right bracketing* of the expression. For instance, the right bracketing of $((x_0(x_1 x_2))x_3)$ is $x_0(x_1(x_2))(x_3)$ and of

$$((x_0 x_1)(x_2 x_3))$$

is $x_0(x_1)(x_2(x_3))$. Give a formal (inductive) definition of right bracketing and prove that there is a one-to-one correspondence between binary and right bracketings.

28. Show that in a right bracketing of $x_0 x_1 \cdots x_n$, there is one and only one opening bracket preceding any x_i, $1 \leq i \leq n$.

29. Associate with a right bracketing of $x_0 x_1 \cdots x_n$ a *bracketing function* $E: \{1, \ldots, n\} \to \{1, \ldots, n\}$ defined as follows: For $1 \leq i \leq n$ there is, by exercise 28, an opening bracket before x_i; let this bracket close following x_j; set $E(i) = j$. Show that E has the following properties: (i) $i \leq E(i)$ for $1 \leq i \leq n$; (ii) $i \leq j \leq E(i)$ imply that $E(j) \leq E(i)$ for $1 \leq i \leq j \leq n$.

*30. Show that (i) and (ii) of exercise 29 characterize bracketing functions.

31. Let E_n denote the set of all bracketing functions defined on $\{1, \ldots, n\}$. For $E, F \in E_n$ set $E \leq F$ iff $E(i) \leq F(i)$ for all i, $1 \leq i \leq n$. Show that $\leq$ is a partial ordering, E_n as a poset is a lattice, and $(E \wedge F)(i) = \inf\{E(i), F(i)\}$.

32. The *semiassociative identity* applied at the place i is a map $\sigma_i: T_n \to T_n$ defined as follows: If $E = \cdots(A(BC))\cdots$, where the first variable in B and C is x_i and x_j, respectively, then $E\sigma_i = \cdots((AB)C)\cdots$; if E is not of such form, then $E\sigma_i = E$. Let X and Y denote the bracketing functions associated with E and $E\sigma_i$, respectively. Show that $X(k) = Y(k)$ for $k \neq i$, $i \le k \le n$, and $Y(i) = j - 1$. Conclude that $X > Y$.

33. Show the converse of exercise 32.

34. For $E, F \in T_n$ define $E < F$ to mean the existence of a sequence $E = X_0$, $X_1, \ldots, X_k = F$, $X_i \in T_n$, $0 \le i \le k$ such that X_{i+1} can be gotten from X_i by some application of the semiassociative law for $0 \le i < k$. Let $E \le F$ mean $E = F$ or $E < F$. Show that $\le$ is a partial ordering. Verify that Figure 2.10 is the diagram of T_3 and T_4. Is the diagram of T_4 optimal? (No)

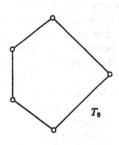 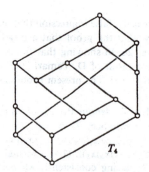

Figure 2.10

*35. Let $X, Y \in E_n$ and $X >- Y$. Let E and F be the binary bracketings associated with X and Y, respectively. Show that $F = E\sigma_i$ for some i.

36. Show that T_n is a lattice for each $n \ge 0$. (In fact, $T_n \cong E_n$.)

3. Some Algebraic Concepts

The purpose of any algebraic theory is the investigation of algebras up to isomorphism. We can introduce two concepts of isomorphism for lattices.

The lattices $\mathfrak{L}_0 = \langle L_0; \le \rangle$, $\mathfrak{L}_1 = \langle L_1; \le \rangle$ are *isomorphic*, (in symbol, $\mathfrak{L}_0 \cong \mathfrak{L}_1$), and the map $\varphi: L_0 \to L_1$ is an *isomorphism* if φ is one-to-one and onto, and

$$a \le b \text{ in } \mathfrak{L}_0 \text{ iff } a\varphi \le b\varphi \text{ in } \mathfrak{L}_1.$$

The lattices $\mathfrak{L}_0 = \langle L; \wedge, \vee \rangle$, $\mathfrak{L}_1 = \langle L_1; \wedge, \vee \rangle$ are *isomorphic* (in symbol, $\mathfrak{L}_0 \cong \mathfrak{L}_1$), and the map $\varphi: L_0 \rightarrow L_1$ is an *isomorphism* if φ is one-to-one and onto, and if

$$(a \wedge b)\varphi = a\varphi \wedge b\varphi$$

and

$$(a \vee b)\varphi = a\varphi \vee b\varphi.$$

It is easy to see that the two concepts coincide under the equivalence of Theorem 1.1. However, when we generalize these to homomorphism concepts, we get various new nonequivalent notions. In order to avoid confusion, each will be given a different name.

From now on we will abandon the precise notation $\langle L; \wedge, \vee \rangle$ and $\langle L; \leq \rangle$ for lattices and posets; we will simply write italic capitals, indicating the underlying sets, unless for some reason we want to be more exact.

Note that the first definition of isomorphism can be applied to any two posets L_0, L_1, thus yielding an isomorphism concept for arbitrary posets. Having this concept of isomorphism, we can restate the content of Lemma 2.1: The diagram of a finite poset determines the poset up to isomorphism.

Let $\mathfrak{C}_n$ denote the set $\{0, \ldots, n-1\}$ ordered by $0 < 1 < 2 < \cdots < n - 1$. Then $\mathfrak{C}_n$ is an n-element chain. If $C = \{x_0, \ldots, x_{n-1}\}$ is an n-element chain, $x_0 < x_1 < \cdots < x_{n-1}$, then $\varphi: i \rightarrow x_i$ is an isomorphism between $\mathfrak{C}_n$ and C. Therefore, the n-element chain is unique up to isomorphism.

The isomorphism of posets generalizes as follows:

The map $\varphi: P_0 \rightarrow P_1$ is an *isotone map* (also called *monotone map*) of the poset P_0 into the poset P_1, if $a \leq b$ in P_0 implies that $a\varphi \leq b\varphi$ in P_1.

A *homomorphism* of the semilattice $\langle S_0; \cdot \rangle$ into the semilattice $\langle S_1; \cdot \rangle$ is a map $\varphi: S_0 \rightarrow S_1$ satisfying $(a \cdot b)\varphi = a\varphi \cdot b\varphi$. Since a lattice $\mathfrak{L} = \langle L; \wedge, \vee \rangle$ is a semilattice both under $\wedge$ and under $\vee$, we get two homomorphism concepts, *meet-homomorphism* ($\wedge$-homomorphism) and *join-homomorphism* ($\vee$-homomorphism). A *homomorphism* is a map that is both a meet-homomorphism and a join-homomorphism. Thus a homomorphism φ of the lattice L_0 into the lattice L_1 is a map of L_0 into L_1 satisfying both $(a \wedge b)\varphi = a\varphi \wedge b\varphi$ and $(a \vee b)\varphi = a\varphi \vee b\varphi$. A one-to-one homomorphism will also be called an *embedding*. (The list of homomorphism concepts will be further extended in Section 6.)

Note that meet-homomorphisms, join-homomorphisms, and (lattice) homomorphisms are all isotone. Let us prove this statement for meet-homomorphisms. If $\varphi: L_0 \rightarrow L_1$, $(a \wedge b)\varphi = a\varphi \wedge b\varphi$ for all $a, b \in L_0$, and

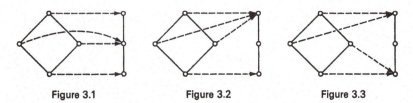

Figure 3.1   Figure 3.2   Figure 3.3

if $x,y \in L_0$, $x \le y$ in L_0, then $x = x \wedge y$; thus $x\varphi = (x \wedge y)\varphi = x\varphi \wedge y\varphi$, and $x\varphi \le y\varphi$ in L_1. Note that the converse does not hold, nor is there any connection between meet- and join-homomorphisms.

Figures 3.1–3.3 are three maps of the four-element lattice L of Figure 2.1 into the three-element chain $\mathfrak{C}_3$. The map of Figure 3.1 is isotone but is neither a meet- nor a join-homomorphism. The map of Figure 3.2 is a join-homomorphism but is not a meet-homomorphism, thus not a homomorphism. The map of Figure 3.3 is a homomorphism.

The second basic algebraic concept is that of a subalgebra:

A *sublattice* $\mathfrak{K} = \langle K; \wedge, \vee \rangle$ of the lattice $\mathfrak{L} = \langle L; \wedge, \vee \rangle$ is defined on a nonvoid subset K of L with the property that $a,b \in K$ implies that $a \wedge b$, $a \vee b \in K$ ($\wedge, \vee$ taken in $\mathfrak{L}$), and the $\wedge$ and the $\vee$ of $\mathfrak{K}$ are restrictions to K of the $\wedge$ and the $\vee$ of $\mathfrak{L}$.

To put this in simpler language, we take a subset K of a lattice L such that K is closed under $\wedge$ and $\vee$. Under the same $\wedge$ and $\vee$, K is a lattice; this is a sublattice of L.

The concept of a lattice as a poset would suggest the following sublattice concept: Take a nonvoid subset K of the lattice L; if the partial ordering of L makes K a lattice, call K a sublattice of L. This concept is different from the previous one and will not be used at all.

Let A_λ, $\lambda \in \Lambda$, be sublattices of L. Then $\bigcap (A_\lambda \mid \lambda \in \Lambda)$ (the set theoretic intersection of A_λ, $\lambda \in \Lambda$) is also closed under $\wedge$ and $\vee$; thus for every $H \subseteq L$, $H \ne \varnothing$, there is a smallest $[H] \subseteq L$ containing H and closed under $\wedge$ and $\vee$. The sublattice $[H]$ is called the *sublattice of L generated by H*, and H is a *generating set* of $[H]$.

The subset K of the lattice L is called *convex*, if $a,b \in K$, $c \in L$, and $a \le c \le b$ imply that $c \in K$. For $a,b \in L$, $a \le b$, the *interval* $[a, b] = \{x \mid a \le x \le b\}$ is an important example of a convex sublattice. For a chain C, $a,b \in C$, $a \le b$, we can also define the *half-open intervals*: $(a, b] = \{x \mid a < x \le b\}$ and $[a, b) = \{x \mid a \le x < b\}$, and the *open interval*: $(a, b) = \{x \mid a < x < b\}$. These are also examples of convex sublattices. A sublattice I of L is an *ideal* if $i \in I$ and $a \in L$ imply that

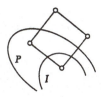

Figure 3.4

$a \wedge i \in I$. An ideal I of L is *proper* if $I \neq L$. A proper ideal I of L is *prime* if $a,b \in L$ and $a \wedge b \in I$ imply that $a \in I$ or $b \in I$. In Figure 3.4, I is an ideal and P is a prime ideal; note that I is not prime.

The concept of a convex sublattice is a typical example of the interplay between the algebraic and order theoretic concepts. We will now examine these concepts more closely.

Since the intersection of any number of convex sublattices (ideals) is a convex sublattice (ideal) unless void, we can define the *convex sublattice generated by a subset H*, and the ideal generated by a subset H of the lattice L, provided that $H \neq \varnothing$. The *ideal generated by a subset H* will be denoted by $(H]$, and if $H = \{a\}$, we write $(a]$ for $(\{a\}]$; we shall call $(a]$ a *principal ideal*.

LEMMA 1. *Let L be a lattice and let H and I be nonvoid subsets of L.*
  (i) *I is an ideal iff $a,b \in I$ implies that $a \vee b \in I$, and $a \in I$, $x \in L$, $x \leq a$ imply that $x \in I$.*
  (ii) *$I = (H]$ iff for all $i \in I$ there exists an integer $n \geq 1$ and there exist $h_0, \ldots, h_{n-1} \in H$ such that $i \leq h_0 \vee \cdots \vee h_{n-1}$.*
 (iii) *For $a \in L$, $(a] = \{x \mid x \leq a\}$.*

PROOF.
  (i) Let I be an ideal. Then $a,b \in I$ implies that $a \vee b \in I$, since I is a sublattice. If $x \leq a \in I$, then $x = x \wedge a \in I$, and the condition of (i) is verified. Conversely, let I satisfy the condition in (i). Let $a,b \in I$. Then $a \vee b \in I$ by definition, and, since $a \wedge b \leq a \in I$, we also have $a \wedge b \in I$; thus I is a sublattice. Finally, if $x \in L$ and $a \in I$, then $a \wedge x \leq a \in I$, thus $a \wedge x \in I$, proving that I is an ideal. ▌
  (ii) Set $I_0 = \{i \mid i \leq h_0 \vee \cdots \vee h_{n-1}$ for $h_0, \ldots, h_{n-1} \in H$ and for some integer $n \geq 1\}$. Using (i), it is clear that I_0 is an ideal, and obviously $H \subseteq I_0$. Finally, if $H \subseteq J$ and J is an ideal, then $I_0 \subseteq J$, and thus I_0 is the smallest ideal containing H; that is, $I = I_0$. ▌
 (iii) This proof is obvious directly, or by application of (ii). ●

Let $I(L)$ denote the set of all ideals of L and let $I_0(L) = I(L) \cup \{\varnothing\}$. We call $I(L)$ the *ideal lattice* and $I_0(L)$ the *augmented ideal lattice* of L.

COROLLARY 2.   *$I(L)$ and $I_0(L)$ are posets under set inclusion, and as posets they are lattices.*

In fact, $I \vee J = (I \cup J]$, if we agree that $(\varnothing] = \varnothing$. From (ii) of Lemma 1 we see that for $I, J \in I(L)$, $x \in I \vee J$ iff $x \le i \vee j$ for some $i \in I$, $j \in J$. It does not matter that we consider only two ideals in this formula. In general,

$$\bigvee (I_\lambda \mid \lambda \in \Lambda) = (\bigcup (I_\lambda \mid \lambda \in \Lambda)];$$

that is, any nonvoid subset of $I(L)$ has a supremum. Combining this formula with Lemma 1(ii), we have:

COROLLARY 3.   *Let I_λ, $\lambda \in \Lambda$, be ideals and let $I = \bigvee (I_\lambda \mid \lambda \in \Lambda)$. Then $i \in I$ iff $i \le j_{\lambda_0} \vee \cdots \vee j_{\lambda_{n-1}}$, for some integer $n \ge 1$ and for some $\lambda_0, \ldots, \lambda_{n-1} \in \Lambda$, $j_{\lambda_i} \in I_{\lambda_i}$.*

Now observe the formulas:

$$(a] \wedge (b] = (a \wedge b], \qquad (a] \vee (b] = (a \vee b].$$

Since $a \ne b$ implies that $(a] \ne (b]$, these formulas yield:

COROLLARY 4.   *L can be embedded in $I(L)$ and also in $I_0(L)$, and $a \to (a]$ is such an embedding.*

Let us connect homomorphisms and ideals (recall that $\mathfrak{C}_2$ denotes the two-element chain with elements 0 and 1).

LEMMA 5.

(i) *I is a proper ideal of L iff there is a join-homomorphism φ of L onto $\mathfrak{C}_2$ such that $I = 0\varphi^{-1}$ (the complete inverse image of 0, that is,*

$$I = \{x \mid x\varphi = 0\}).$$

(ii) *I is a prime ideal of L iff there is a homomorphism φ of L onto $\mathfrak{C}_2$ with $I = 0\varphi^{-1}$.*

PROOF.

(i) Let I be a proper ideal and define φ by $x\varphi = 0$ if $x \in I$, $x\varphi = 1$ if $x \notin I$; obviously, this φ is a join-homomorphism. Conversely, if φ is a join-homomorphism of L onto $\mathfrak{C}_2$ and $I = 0\varphi^{-1}$, then for $a,b \in I$, we have $a\varphi = b\varphi = 0$; thus $(a \vee b)\varphi = a\varphi \vee b\varphi = 0 \vee 0 = 0$, that is, $a \vee b \in I$. If $a \in I$, $x \in L$, $x \leq a$, then $x\varphi \leq a\varphi = 0$, that is, $x\varphi = 0$; thus $x \in I$. Finally, φ is onto, therefore $I \neq L$. ▶

(ii) If I is prime, take the φ constructed in the proof of (i) and note that φ can violate the property of being a homomorphism only with $a,b \notin I$. However, since I is prime, $a \wedge b \notin I$; consequently $(a \wedge b)\varphi = 1 = a\varphi \wedge b\varphi$, and so φ is a homomorphism. Conversely, let φ be a homomorphism of L onto $\mathfrak{C}_2$ and let $I = 0\varphi^{-1}$. If $a,b \notin I$, then $a\varphi = b\varphi = 1$, thus $(a \wedge b)\varphi = a\varphi \wedge b\varphi = 1$, and therefore $a \wedge b \notin I$; I is prime. ●

By dualizing, we get the concepts of *dual ideal* (also called *filter*[1]), *principal dual ideal*, $[a)$ (*principal filter*), the *dual ideal* $[H)$ generated by H, *proper dual ideal*, *prime dual ideal* (*prime filter*, or *ultra filter*), the lattice $\mathscr{D}(L)$ of dual ideals ordered by set inclusion, and $\mathscr{D}_0(L) = \mathscr{D}(L) \cup \{\varnothing\}$ ordered by set inclusion. Note that in $\mathscr{D}(L)$ (and $\mathscr{D}_0(L)$) the largest element is L; if L has 0 and 1, then $L = [0)$ is the largest and $\{1\} = [1)$ is the smallest element of $\mathscr{D}(L)$. Furthermore, for $a,b \in L$ we have

$$[a) \wedge [b) = [a \vee b), \quad \text{and} \quad [a) \vee [b) = [a \wedge b).$$

LEMMA 6.   *Let I be an ideal and let D be a dual ideal. If $I \cap D \neq \varnothing$, then $I \cap D$ is a convex sublattice, and every convex sublattice can be expressed in this form in one and only one way.*

PROOF.   The first statement is obvious. To prove the second, let C be a convex sublattice and set $I = (C]$, $D = [C)$. Then $C \subseteq I \cap D$. If $t \in I \cap D$, then $t \in I$, and thus by (ii) of Lemma 1, $t \leq c$ for some $c \in C$; also, $t \in D$; therefore, by the dual of (ii) of Lemma 1, $t \geq d$ for some $d \in C$. This implies that $t \in C$ since C is convex, and so $C = I \cap D$.

Suppose now that C has another representation, $C = I_1 \cap D_1$. Since $C \subseteq I_1$, we have $(C] \subseteq I_1$. Let $a \in I_1$ and let c be an arbitrary element of C. Then $a \vee c \in I_1$ and $a \vee c \geq c \in D_1$, so $a \vee c \in D_1$, thus $a \vee c \in I_1 \cap D_1 = C$. Finally, $a \leq a \vee c \in C$; therefore, $a \in (C]$. This shows that $I_1 = (C]$. The dual argument shows that $D_1 = [C)$. Hence the uniqueness of such representations. ●

---

[1] In the literature, filter means one of the following four concepts: (i) dual ideal; (ii) proper dual ideal; (iii) dual ideal of the lattice of all subsets of a set; (iv) proper dual ideal of the lattice of all subsets of a set. Further variants allow the empty set as a filter under (i) or (ii).

An *equivalence relation* Θ (that is, a reflexive, symmetric, and transitive binary relation) on a lattice L is called a *congruence relation* of L if $a_0 \equiv b_0(\Theta)$ and $a_1 \equiv b_1(\Theta)$ imply that $a_0 \wedge a_1 \equiv b_0 \wedge b_1(\Theta)$ and $a_0 \vee a_1 \equiv b_0 \vee b_1(\Theta)$ (*Substitution Property*). Trivial examples are ω, ι, defined by $x \equiv y(\omega)$ iff $x = y$; $x \equiv y(\iota)$ for all x and y. For $a \in L$, we write $[a]\Theta$ for the *congruence class* containing a, that is, $[a]\Theta = \{x \mid x \equiv a(\Theta)\}$.

LEMMA 7. *Let Θ be a congruence relation of L. Then for every $a \in L$, $[a]\Theta$ is a convex sublattice.*

PROOF. Let $x,y \in [a]\Theta$; then $x \equiv a(\Theta)$ and $y \equiv a(\Theta)$. Therefore, $x \wedge y \equiv a \wedge a = a(\Theta)$, and $x \vee y \equiv a \vee a = a(\Theta)$, proving that $[a]\Theta$ is a sublattice. If $x \leq t \leq y$, $x,y \in [a]\Theta$, then $x \equiv a(\Theta)$ and $y \equiv a(\Theta)$. Therefore, $t = t \wedge y \equiv t \wedge a(\Theta)$, and $t = t \vee x \equiv (t \wedge a) \vee x \equiv (t \wedge a) \vee a = a(\Theta)$, proving that $[a]\Theta$ is convex. ●

Sometimes a long computation is required to prove that a given binary relation is a congruence relation. Such computations are often facilitated by the following lemma (G. Grätzer and E. T. Schmidt [1958e]):

LEMMA 8. *A reflexive and symmetric binary relation Θ on a lattice L is a congruence relation iff the following three properties are satisfied for $x,y,z,t \in L$:*

(i) $x \equiv y(\Theta)$ iff $x \wedge y \equiv x \vee y(\Theta)$.
(ii) $x \leq y \leq z$, $x \equiv y(\Theta)$, and $y \equiv z(\Theta)$ imply that $x \equiv z(\Theta)$.
(iii) $x \leq y$ and $x \equiv y(\Theta)$ imply that $x \wedge t \equiv y \wedge t(\Theta)$ and $x \vee t \equiv y \vee t(\Theta)$.

PROOF. The "only if" part being trivial, assume now that a symmetric and reflexive binary relation Θ satisfies conditions (i)–(iii). Let $b,c \in [a, d]$ and $a \equiv d(\Theta)$; we claim that $b \equiv c(\Theta)$. Indeed, $a \equiv d(\Theta)$ and $a \leq d$ imply by (iii) that $b \wedge c = a \vee (b \wedge c) \equiv d \vee (b \wedge c) = d(\Theta)$. Now $b \wedge c \leq d$ and (iii) imply that $b \wedge c = (b \wedge c) \wedge (b \vee c) \equiv d \wedge (b \vee c) = b \vee c(\Theta)$; thus by (i), $b \equiv c(\Theta)$.

To prove that Θ is transitive, let $x \equiv y(\Theta)$ and $y \equiv z(\Theta)$. Then by (i), $x \wedge y \equiv x \vee y(\Theta)$, and by (iii), $y \vee z = (y \vee z) \vee (y \wedge x) \equiv (y \vee z) \vee (y \vee x) = x \vee y \vee z(\Theta)$, and similarly, $x \wedge y \wedge z \equiv y \wedge z(\Theta)$. Therefore, $x \wedge y \wedge z \equiv y \wedge z \equiv y \vee z \equiv x \vee y \vee z(\Theta)$ and $x \wedge y \wedge z \leq y \wedge z \leq y \vee z \leq x \vee y \vee z$. Thus, applying (ii) twice, we get $x \wedge y \wedge z \equiv x \vee y \vee z(\Theta)$. Now we apply the statement of the previous paragraph with $a = x \wedge y \wedge z$, $b = x$, $c = z$, $d = x \vee y \vee z$ to conclude that $x \equiv z(\Theta)$.

Let $x \equiv y(\Theta)$; we claim that $x \vee t \equiv y \vee t(\Theta)$. Indeed, $x \wedge y \equiv x \vee y(\Theta)$ by (i); thus by (iii), $(x \wedge y) \vee t \equiv x \vee y \vee t(\Theta)$. Since $x \vee t$, $y \vee t \in [(x \wedge y) \vee t, \ x \vee y \vee t]$, we conclude that $x \vee t \equiv y \vee t(\Theta)$.

To prove the Substitution Property for $\vee$, let $x_0 \equiv y_0(\Theta)$ and $x_1 \equiv y_1(\Theta)$. Then $x_0 \vee x_1 \equiv x_0 \vee y_1 \equiv y_0 \vee y_1(\Theta)$, implying that $x_0 \vee x_1 \equiv y_0 \vee y_1(\Theta)$ since Θ is transitive. The Substitution Property for $\wedge$ is similarly proved. ●

Let $C(L)$ denote the set of all congruence relations on L partially ordered by set inclusion (remember that every $\Theta \in C(L)$ is a subset of L^2). As a first application of Lemma 8 we prove

THEOREM 9. $C(L)$ *is a lattice. For* $\Theta, \Phi \in C(L)$, $\Theta \wedge \Phi = \Theta \cap \Phi$. *The join,* $\Theta \vee \Phi$, *can be described as follows:*

$x \equiv y (\Theta \vee \Phi)$ iff there is a sequence $z_0 = x \wedge y, z_1, \ldots, z_{n-1} = x \vee y$ of elements of L such that $z_0 \leq z_1 \leq \cdots \leq z_{n-1}$ and for each i, $0 \leq i < n - 1$, $z_i \equiv z_{i+1}(\Theta)$ or $z_i \equiv z_{i+1}(\Phi)$.

REMARK. $C(L)$ is called the *congruence lattice* of L. Observe that $C(L)$ is a sublattice of $E(L)$ (exercises 45, 46); that is, the join and meet of congruence relations as congruence relations and as equivalence relations coincide.

PROOF.   The first statement is obvious. To prove the second statement, let Ψ be the binary relation described in Theorem 9. Then $\Theta \subseteq \Psi$ and $\Phi \subseteq \Psi$ are obvious. If Γ is a congruence relation, $\Theta \subseteq \Gamma$, $\Phi \subseteq \Gamma$, and $x \equiv y(\Psi)$, then for each i, either $z_i \equiv z_{i+1}(\Theta)$ or $z_i \equiv z_{i+1}(\Phi)$; thus $z_i \equiv z_{i+1}(\Gamma)$. By the transitivity of Γ, $x \wedge y \equiv x \vee y(\Gamma)$; thus $x \equiv y(\Gamma)$. Therefore, $\Psi \subseteq \Gamma$. This shows that if Ψ is a congruence relation, then $\Psi = \Theta \vee \Phi$.

Ψ is obviously reflexive and symmetric. If $x \leq y \leq z$, $x \equiv y(\Psi)$, and $y \equiv z(\Psi)$, then $x \equiv z(\Psi)$ is established by putting together the sequences showing $x \equiv y(\Psi)$ and $y \equiv z(\Psi)$. Let $x \equiv y(\Psi)$, $x \leq y$, with z_i, $0 \leq i < n$ establishing this, and $t \in L$. Then $x \wedge t \equiv y \wedge t(\Psi)$, $x \vee t \equiv y \vee t(\Psi)$ can be shown with the sequences $z_i \wedge t$, $0 \leq i < n$, $z_i \vee t$, $0 \leq i < n$, respectively. Thus (i)–(iii) of Lemma 8 hold for Ψ, and we conclude that Ψ is a congruence relation. ●

Homomorphisms and congruence relations express two sides of the same phenomenon. To establish this fact we first define quotient lattices (also called factor lattices). Let L be a lattice and let Θ be a congruence relation on L. Let L/Θ denote the set of blocks of the partition of L induced by Θ, that is,

$$L/\Theta = \{[a]\Theta \mid a \in L\}.$$

Set

$$[a]\Theta \wedge [b]\Theta = [a \wedge b]\Theta$$

and

$$[a]\Theta \vee [b]\Theta = [a \vee b]\Theta.$$

This defines $\wedge$ and $\vee$ on L/Θ. Indeed, if $[a]\Theta = [a_1]\Theta$ and $[b]\Theta = [b_1]\Theta$, then $a \equiv a_1(\Theta)$ and $b \equiv b_1(\Theta)$; therefore, $a \wedge b \equiv a_1 \wedge b_1(\Theta)$, that is, $[a \wedge b]\Theta = [a_1 \wedge b_1]\Theta$. Thus $\wedge$ and (dually) $\vee$ are well defined on L/Θ. The lattice axioms are easily verified. The lattice L/Θ is the *quotient lattice* of L modulo Θ.

LEMMA 10. *The map*

$$\varphi_\Theta \colon x \to [x]\Theta \qquad (x \in L)$$

is a homomorphism of L onto L/Θ.

REMARK.   The lattice K is a *homomorphic image* of the lattice L if there is a homomorphism of L onto K. Lemma 10 states that any quotient lattice is a homomorphic image.

PROOF.   The proof is trivial. ●

THEOREM 11 (THE HOMOMORPHISM THEOREM).   *Every homomorphic image of a lattice L is isomorphic to a suitable quotient lattice of L. In fact, if $\varphi \colon L \to L_1$ is a homomorphism of L onto L_1 and if Θ is the congruence relation of L defined by $x \equiv y(\Theta)$ iff $x\varphi = y\varphi$, then*

$$L/\Theta \cong L_1;$$

an isomorphism (see Figure 3.5) is given by

$$\psi \colon [x]\Theta \to x\varphi, \qquad x \in L.$$

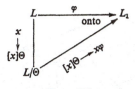

Figure 3.5

PROOF. It is easy to check that Θ is a congruence relation. To prove that ψ is an isomorphism we have to check that ψ (i) is well defined, (ii) is one-to-one, (iii) is onto, and (iv) preserves the operations.

(i) Let $[x]\Theta = [y]\Theta$. Then $x \equiv y(\Theta)$; thus $x\varphi = y\varphi$, that is,

$$([x]\Theta)\psi = ([y]\Theta)\psi.$$

(ii) Let $([x]\Theta)\psi = ([y]\Theta)\psi$, that is, $x\varphi = y\varphi$. Then $x \equiv y(\Theta)$, and so $[x]\Theta = [y]\Theta$.

(iii) Let $a \in L_1$. Since φ is onto, there is an $x \in L$ with $x\varphi = a$. Thus $([x]\Theta)\psi = a$.

(iv) $([x]\Theta \wedge [y]\Theta)\psi = ([x \wedge y]\Theta)\psi = (x \wedge y)\varphi = x\varphi \wedge y\varphi$

$$= ([x]\Theta)\psi \wedge ([y]\Theta)\psi.$$

The computation for $\vee$ is identical. ●

The final algebraic concept introduced in this section is that of direct product. Let L, K be lattices and form the set $L \times K$ of all ordered pairs $\langle a, b \rangle$ with $a \in L$, $b \in K$. Define $\wedge$ and $\vee$ in $L \times K$ "componentwise":

$$\langle a, b \rangle \wedge \langle a_1, b_1 \rangle = \langle a \wedge a_1, b \wedge b_1 \rangle$$

$$\langle a, b \rangle \vee \langle a_1, b_1 \rangle = \langle a \vee a_1, b \vee b_1 \rangle$$

This makes $L \times K$ into a lattice, called the *direct product* of L and K (for an example, see Figure 3.6).

LEMMA 12. *Let* L, L_1, K, K_1 *be lattices,* $L \cong L_1$, $K \cong K_1$. *Then*

$$L \times K \cong L_1 \times K_1 \cong K_1 \times L_1.$$

REMARK. This means that $L \times K$ is determined up to isomorphism if we know L and K up to isomorphism, and the direct product is determined up to isomorphism by the factors, the order in which they are given being irrelevant.

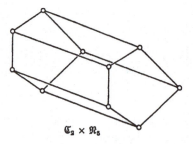

$\mathfrak{C}_2 \times \mathfrak{N}_5$

Figure 3.6

PROOF. Let $\varphi: L \to L_1$ and $\psi: K \to K_1$ be isomorphisms and for $a \in L$, $b \in K$ define $\langle a, b \rangle \chi = \langle a\varphi, b\psi \rangle$. Then $\chi: L \times K \to L_1 \times K_1$ is an isomorphism. Of course, $L_1 \times K_1 \cong K_1 \times L_1$ is proved by showing that $\langle a, b \rangle \to \langle b, a \rangle$ $(a \in L_1, b \in K_1)$ is an isomorphism. ●

If L_i, $i \in I$, is a family of lattices, again we first form the Cartesian product of the sets $\prod (L_i \mid i \in I)$, which is defined as the set of all functions $f: I \to \bigcup (L_i \mid i \in I)$ such that $f(i) \in L_i$, for all $i \in I$. We then define $\wedge$ and $\vee$ "componentwise"; that is, $f \wedge g = h$, $f \vee g = k$ means:

$$f(i) \wedge g(i) = h(i), \qquad f(i) \vee g(i) = k(i), \qquad \text{for all } i \in I.$$

The resulting lattice is the direct product $\prod (L_i \mid i \in I)$. If $L_i = L$ for all $i \in I$, we get the *direct power* L^I. Letting n denote the set $\{0, \ldots, n-1\}$, L^n is $\underbrace{((L \times L) \times \cdots) \times L}_{n\text{-times}}$ (at least up to isomorphism). In particular, if we identify $f: 2 \to L$ with $\langle f(0), f(1) \rangle$, then we get $L^2 = L \times L$.

A very important property of direct products is:

THEOREM 13. *Let L and K be lattices, let Θ be a congruence relation of L, and let Φ be a congruence relation of K. Define the relation $\Theta \times \Phi$ on $L \times K$ by*

$$\langle a, b \rangle \equiv \langle c, d \rangle \ (\Theta \times \Phi) \ \textit{iff } a \equiv c(\Theta) \ \textit{and } b \equiv d(\Phi).$$

Then $\Theta \times \Phi$ is a congruence relation on $L \times K$. Conversely, every congruence relation of $L \times K$ is of this form.

PROOF. The first statement is obvious. Now let Ψ be a congruence relation on $L \times K$. For $a, b \in L$ define $a \equiv b(\Theta)$ if $\langle a, c \rangle \equiv \langle b, c \rangle (\Psi)$ for some $c \in K$. Let $d \in K$. Joining both sides with $\langle a \wedge b, d \rangle$ and then meeting with $\langle a \vee b, d \rangle$, we get $\langle a, d \rangle \equiv \langle b, d \rangle (\Psi)$; thus $\langle a, c \rangle \equiv \langle b, c \rangle$ for *some*

$c \in K$ is equivalent to $\langle a, c \rangle \equiv \langle b, c \rangle$ for *all* $c \in K$. Similarly, define for $a, b \in K$, $a \equiv b(\Phi)$ iff $\langle c, a \rangle \equiv \langle c, b \rangle(\Psi)$ for any/for all $c \in L$. It is easily seen that Θ and Φ are congruences. Let $\langle a, b \rangle \equiv \langle c, d \rangle(\Theta \times \Phi)$; then $\langle a, x \rangle \equiv \langle c, x \rangle(\Psi)$, $\langle y, b \rangle \equiv \langle y, d \rangle(\Psi)$. Joining the two with $y = a \wedge c$ and $x = b \wedge d$, we get $\langle a, b \rangle \equiv \langle c, d \rangle(\Psi)$. Finally, let $\langle a, b \rangle \equiv \langle c, d \rangle(\Psi)$. Meeting with $\langle a \vee c, b \wedge d \rangle$, we get $\langle a, b \wedge d \rangle \equiv \langle c, b \wedge d \rangle(\Psi)$; therefore, $a \equiv c(\Theta)$. Similarly, $b \equiv d(\Phi)$, and so

$$\langle a, b \rangle \equiv \langle c, d \rangle(\Theta \times \Phi),$$

proving that $\Psi = \Theta \times \Phi$. ●

In conclusion, we introduce a nonalgebraic concept to balance the false impression that might have been created in this section that all lattice theoretic concepts are algebraic.

A lattice L is called *complete* if $\bigwedge H$ and $\bigvee H$ exist for *any subset* $H \subseteq L$. The concept is self-dual, and half of the hypothesis is redundant.

LEMMA 14. *Let P be a poset in which $\bigwedge H$ exists for all $H \subseteq P$. Then P is a complete lattice.*

PROOF. For $H \subseteq P$, let K be the set of all upper bounds of H. By hypothesis, $\bigwedge K$ exists; set $a = \bigwedge K$. If $h \in H$, then $h \leq k$ for all $k \in K$; therefore $h \leq a$, and $a \in K$. Thus, a is the smallest member of K, that is, $a = \bigvee H$. ●

Lemma 14 can be applied to $I_0(L)$ and $C(L)$. For a further application, let Sub (L) denote the set of all subsets A of L closed under $\wedge$ and $\vee$ partially ordered under set inclusion. In other words, if $A \in$ Sub (L), $A \neq \emptyset$, then A is a sublattice of L. Obviously, Sub (L) is closed under arbitrary intersections.

COROLLARY 15. *$I_0(L)$ and $C(L)$ are complete lattices. If L has a smallest element, $I(L)$ is a complete lattice. The lattice Sub (L) is a complete lattice.*

Exercises

1. Formalize and prove the equivalence of the two isomorphism concepts.
2. Let $\varphi: L_0 \to L_1$, $\psi: L_1 \to L_0$ be (lattice) homomorphisms. Show that if $\varphi\psi$ is the identity map on L_0 and $\psi\varphi$ is the identity map on L_1, then φ is an isomorphism and $\psi = \varphi^{-1}$. Furthermore, prove that if $\varphi: L_0 \to L_1$,

$\psi: L_1 \to L_2$ are isomorphisms, then so are φ^{-1} (the *inverse map*, $\varphi^{-1}: L_1 \to L_0$) and $\varphi\psi$.

3. A one-to-one and onto homomorphism is an isomorphism. Is a one-to-one and onto isotone map an isomorphism?

4. Find a general construction of meet-homomorphisms and join-homomorphisms that are not homomorphisms.

5. Find a subset H of a lattice L such that H is not a sublattice of L but H is a lattice under the partial ordering of L restricted to H.

6. Show that a lattice is a chain iff every nonvoid subset of L is a sublattice.

7. Prove that a sublattice generated by two distinct elements has two or four elements.

*8. Find an infinite lattice generated by three elements.

9. Verify that a nonempty subset I of a lattice L is an ideal iff, for $a,b \in L$, $a \vee b \in I$ is equivalent to $a,b \in I$.

10. Prove that if L is finite, then L and $I(L)$ are isomorphic. How about $I_0(L)$?

11. Is there an infinite lattice L such that $L \cong I(L)$, but not every ideal is principal? (There is no such lattice; see D. Higgs [1971].)

12. Prove the completeness of $I_0(L)$ without using Lemma 14.

13. Prove that $C(L)$ is complete by showing the following description of infinite joins: Let $H \subseteq C(L)$, $\Phi = \bigvee H$. Then $x \equiv y(\Phi)$ iff there exists a finite sequence $x \wedge y = z_0, z_1, \ldots, z_{n-1} = x \vee y$, such that $z_0 \leq z_1 \leq \cdots \leq z_{n-1}$, and $z_i \equiv z_{i+1}(\Theta_i)$ for $0 \leq i < n$ and for some $\Theta_i \in H$.

14. Let $\varphi: L \to K$ be an onto homomorphism, let I be an ideal of L, and let J be an ideal of K. Show that $I\varphi$ is an ideal of K, and $J\varphi^{-1} = \{a \mid a \in L, a\varphi \in J\}$ is an ideal of L.

15. Is the image $P\varphi$ of a prime ideal under a homomorphism φ prime again?

16. Show that the complete inverse image $P\varphi^{-1}$ of a prime ideal P under an onto lattice homomorphism φ is prime again.

17. Show that an ideal P is a prime ideal of L iff $L - P$ is a dual ideal—in fact, a prime dual ideal.

18. Let H be a subset of the lattice L such that $a,b \in H$ implies that $a \vee b \in H$. Then

$$(H] = \{t \mid t \leq h \text{ for some } h \in H\};$$

that is,

$$(H] = \bigcup ((h] \mid h \in H).$$

19. Show that if K is a sublattice of L, then K is isomorphic to a sublattice of $I(L)$.

*20. Verify that the converse of exercise 19 is false even for some finite K.

21. Prove that if $\mathfrak{N}_5$ (see Figure 2.2) is isomorphic to a sublattice of $I(L)$, then $\mathfrak{N}_5$ is isomorphic to a sublattice of L.

22. Find a lattice L and a convex sublattice C of L that cannot be represented as $[a]\Theta$ for any congruence relation Θ of L.

23. State and prove an analogue of Lemma 8 for join-congruence relations.

24. Describe $\Theta \vee \Phi$ for join-congruences.

*25. Find a lattice L such that $L \cong L/\Theta$ for all $\Theta \neq \omega$, and there are infinitely many $\Theta \neq \omega$.

26. Describe the congruence lattice of $\mathfrak{N}_5$.

27. Describe the congruence lattice of an n-element chain.

28. Describe the congruence lattice of the lattice of Figure 2.3, and list all quotient lattices.

29. Construct a lattice that has exactly three congruence relations.

30. Construct infinitely many lattices L such that each lattice is isomorphic to its congruence lattice.

*31. Can an L in exercise 30 be infinite?

32. Generalize Lemma 12 to the direct product of more than two lattices.

33. Show that $\mathfrak{N}_5 \cong L \times K$ implies that L or K has only one element.

34. Show that $I(L)$ is *conditionally complete*: Every nonvoid set H with an upper bound has a supremum, and dually. $I(L)$ is complete iff L has a smallest element.

35. Let L and K be lattices, let $\varphi \colon L \to K$ be an onto homomorphism, and let K have a smallest element 0. Then $0\varphi^{-1}$ is called the *ideal kernel* of the homomorphism φ. Show that the ideal kernel of a homomorphism is an ideal of L.

36. Find an ideal that is the ideal kernel of no homomorphism.

37. Find an ideal that is the kernel of more than one (infinitely many) homomorphisms.

38. Prove that every ideal of a lattice L is prime iff L is a chain.

39. Under what conditions is $L \times K$ planar?

40. Let L and K be lattices and let $\varphi \colon L \to K$ be one-to-one and onto satisfying $\{(a \wedge b)\varphi, (a \vee b)\varphi\} = \{a\varphi \wedge b\varphi, a\varphi \vee b\varphi\}$ for all $a,b \in L$. Let $A \in \text{Sub}(L)$. Show that $A\varphi \in \text{Sub}(K)$; that in fact $A \to A\varphi$ is a (lattice) isomorphism between $\text{Sub}(L)$ and $\text{Sub}(K)$.

41. Prove the converse of exercise 40.

42. Generalize Theorem 13 to finitely many lattices.

43. Show that the first part of Theorem 13 holds for any number of lattices but that the second part does not.

44. Show that the second statement of Theorem 13 fails for (Abelian) groups.

45. For a set X, let $E(X)$ denote the set of all equivalence relations on X partially ordered under set inclusion. Show that $E(X)$ is a (complete) lattice.

46. Show that $C(L)$ is a sublattice of $E(L)$.

4. Polynomials, Identities, and Inequalities

From variables $x_0, x_1, \ldots, x_n, \ldots$, we can form polynomials in the usual manner using $\wedge$, $\vee$, and, of course, parentheses. Examples of polynomials are: $x_0, x_3, x_0 \vee x_0, (x_0 \wedge x_2) \vee (x_3 \wedge x_0), (x_0 \wedge x_1) \vee$ $(x_0 \wedge x_2) \vee (x_1 \wedge x_2)$. A formal definition is:

DEFINITION 1. *The set* $\mathbf{P}^{(n)}$ *of n-ary lattice polynomials* is *the smallest set satisfying* (i) *and* (ii):
(i) $x_i \in \mathbf{P}^{(n)}, 0 \leq i < n$.
(ii) *If* $p, q \in \mathbf{P}^{(n)}$, *then* $(p \wedge q), (p \vee q) \in \mathbf{P}^{(n)}$.

REMARK. We shall omit the outside parentheses and write $p_1 \vee \cdots \vee p_n$ for $(\cdots (p_1 \vee p_2) \cdots \vee p_n)$, and the same for $\wedge$. Thus we write $x_0 \vee x_1$ for $(x_0 \vee x_1)$ and $x_0 \vee x_1 \vee x_2$ for $((x_0 \vee x_1) \vee x_2)$.

By Definition 1, a polynomial is just a sequence of symbols. It is defined because in terms of such a sequence of symbols we can define a function on any lattice:

DEFINITION 2. *An n-ary polynomial* p *defines a function in n-variables* (*a polynomial function, or simply, a polynomial*) *on a lattice L by the following rules* $(a_0, \ldots, a_{n-1} \in L)$:
(i) *If* $p = x_i, p(a_0, \ldots, a_{n-1}) = a_i, 0 \leq i < n$.
(ii) *If* $p(a_0, \ldots, a_{n-1}) = a, q(a_0, \ldots, a_{n-1}) = b$, *and* $p \wedge q = r, p \vee q = t$, *then* $r(a_0, \ldots, a_{n-1}) = a \wedge b$ *and* $t(a_0, \ldots, a_{n-1}) = a \vee b$.

Thus if $p = (x_0 \wedge x_1) \vee (x_2 \vee x_1)$, then $p(a, b, c) = (a \wedge b) \vee (c \vee b)$ $= b \vee c$. Definitions 1 and 2 are quite formal but their meaning is very simple.

A *polynomial* is an *n*-ary polynomial for some *n*. We will also use $x, y, z, \ldots$ instead of the x_i.

Note that if p is a *unary* ($n = 1$) lattice polynomial, then $p(a) = a$ for any $a \in L$. If p is *binary*, then $p(a, b) = a$, or b, or $a \wedge b$, or $a \vee b$.

We shall prove statements on polynomials by induction on the *rank*. The rank of x_i is 1; that of $p \wedge q$ (and of $p \vee q$) is the sum of the ranks of p and q.

Now we are in a position to describe $[H]$, the sublattice generated by H:

LEMMA 3. $a \in [H]$ *iff* $a = p(h_0, \ldots, h_{n-1})$ *for some integer* $n \geq 1$, *for some n-ary polynomial* p, *and for* $h_0, \ldots, h_{n-1} \in H$.

PROOF. First we must show that if $a = p(h_0, \ldots, h_{n-1})$, $h_i \in H$, then $a \in [H]$, which can be easily accomplished by induction on the rank of p.

Then we form the set $\{a \mid a = p(h_0, \ldots, h_{n-1}), n \geq 1, h_i \in H\}$ and observe that it contains H and that it is closed under $\wedge$ and $\vee$. Since it is contained in $[H]$, it has to equal $[H]$. ●

COROLLARY 4. $|[H]| \leq |H| + \aleph_0$.

PROOF.   By Lemma 3, every element of $[H]$ can be associated with a finite sequence of elements of $H \cup \{(,), \wedge, \vee\}$, and there are no more than $|H| + \aleph_0$ such sequences. ●

DEFINITION 5.   *A lattice* identity (inequality) *is an expression of the form* $p = q$ $(p \leq q)$, *where p and q are polynomials. An* identity (inequality) *holds* in the lattice L *if* $p(a_0, a_1, \ldots) = q(a_0, a_1, \ldots)$ $(p(a_0, a_1, \ldots) \leq q(a_0, a_1, \ldots))$ *holds for any* $a_0, a_1, \ldots \in L$.

An identity $p = q$ is equivalent to the two inequalities $p \leq q$ and $q \leq p$, and the inequality $p \leq q$ is equivalent to the identity $p \vee q = q$ (and to $p \wedge q = p$). Frequently, the validity of identities is shown by verifying that the two inequalities hold.

One of the most basic properties of polynomials is:

LEMMA 6.   *A polynomial* (*function*) *p is isotone; that is, if* $a_0 \leq b_0$, $a_1 \leq b_1, \ldots$, *then* $p(a_0, a_1, \ldots) \leq p(b_0, b_1, \ldots)$. *Furthermore,*

$$x_0 \wedge \cdots \wedge x_{n-1} \leq p(x_0, \ldots, x_{n-1}) \leq x_0 \vee \cdots \vee x_{n-1}.$$

PROOF.   We prove the first statement by induction on the rank of p. The first statement is certainly true for $p = x_i$. Suppose that it is true for q and r and that $p = q \wedge r$. Then

$$
\begin{aligned}
p(a_0, \ldots) \wedge p(b_0, \ldots) \\
&= (q(a_0, \ldots) \wedge r(a_0, \ldots)) \wedge (q(b_0, \ldots) \wedge r(b_0, \ldots)) \\
&= (q(a_0, \ldots) \wedge q(b_0, \ldots)) \wedge (r(a_0, \ldots) \wedge r(b_0, \ldots)) \\
&= q(a_0, \ldots) \wedge r(a_0, \ldots) = p(a_0, \ldots);
\end{aligned}
$$

thus $p(a_0, \ldots) \leq p(b_0, \ldots)$. The proof is similar for $p = q \vee r$. Since $x_0 \wedge \cdots \wedge x_{n-1} \leq x_i \leq x_0 \vee \cdots \vee x_{n-1}$ for $0 \leq i \leq n - 1$, using the idempotency of $\wedge$ and $\vee$, we obtain:

$$x_0 \wedge \cdots \wedge x_{n-1} = p(x_0 \wedge \cdots \wedge x_{n-1}, \ldots, x_0 \wedge \cdots \wedge x_{n-1})$$
$$\leq p(x_0, x_1, \ldots, x_{n-1})$$
$$\leq p(x_0 \vee \cdots \vee x_{n-1}, \ldots, x_0 \vee \cdots \vee x_{n-1})$$
$$= x_0 \vee x_1 \vee \cdots \vee x_{n-1},$$

proving the second statement. ●

A simple application is:

LEMMA 7. *Let $p_i = q_i$, $0 \leq i < n$, be lattice identities. Then there is a single identity $p = q$ such that all $p_i = q_i$, $0 \leq i < n$, hold in a lattice L iff $p = q$ holds in L.*

PROOF. Let us take two identities, $p_0 = q_0$ and $p_1 = q_1$. Suppose that all polynomials are n-ary and consider the identity:

(N) $p_0(x_0, \ldots, x_{n-1}) \wedge p_1(x_n, \ldots, x_{2n-1})$
$$= q_0(x_0, \ldots, x_{n-1}) \wedge q_1(x_n, \ldots, x_{2n-1}).$$

It is obvious that if $p_0 = q_0$, $p_1 = q_1$ hold in L, then (N) holds in L. Now let (N) hold in L and let $a_0, \ldots, a_{n-1} \in L$. Substitute $x_0 = a_0, \ldots, x_{n-1} = a_{n-1}$, $x_n = \cdots = x_{2n-1} = a_0 \vee \cdots \vee a_{n-1} = a$. By Lemma 6,

$$p_0(a_0, \ldots, a_{n-1}) \leq p_0(a, \ldots, a) = a = p_1(a, \ldots, a)$$

and similarly for q_0, q_1; thus (N) yields $p_0(a_0, \ldots, a_{n-1}) = q_0(a_0, \ldots, a_{n-1})$. The second identity is derived similarly from (N). The general proof is similar. ●

The most important (and, in fact, characteristic) properties of identities are given by

LEMMA 8. *Identities are preserved under the formation of sublattices, homomorphic images, direct products, and ideal lattices.*

PROOF. Let the polynomials p, q both be n-ary and let $p = q$ hold in L. If L_1 is a sublattice of L, then $p = q$ obviously holds in L_1. Let $\varphi : L \to K$ be an onto homomorphism. A simple induction shows that

$$p(a_0, \ldots, a_{n-1})\varphi = p(a_0\varphi, \ldots, a_{n-1}\varphi)$$

and proves the similar formula for q. Therefore,

$$p(a_0\varphi, \ldots, a_{n-1}\varphi) = p(a_0, \ldots, a_{n-1})\varphi$$
$$= q(a_0, \ldots, a_{n-1})\varphi = q(a_0\varphi, \ldots, a_{n-1}\varphi),$$

and so $p = q$ holds in K. The statement for direct products is also obvious.

The last statement is an obvious corollary of the following formula: Let p be an n-ary polynomial and let $I_0, \ldots, I_{n-1}$ be ideals of L. Then $I_0, \ldots, I_{n-1} \in I(L)$; thus we can substitute the I_j into p: $p(I_0, \ldots, I_{n-1})$ is also in $I(L)$, that is, an ideal of L. This ideal can be described by a simple formula:

$$p(I_0, \ldots, I_{n-1}) = \{x \mid x \leq p(i_0, \ldots, i_{n-1}),$$
$$\text{for some } i_0 \in I_0, \ldots, i_{n-1} \in I_{n-1}\}.$$

This follows easily from Lemma 6 and the formula in Section 3 describing $I \vee J$, by induction on the rank of p. ●

Now we list a few important inequalities:

LEMMA 9.   *The following inequalities hold in any lattice:*
 (i) $(x \wedge y) \vee (x \wedge z) \leq x \wedge (y \vee z)$
 (ii) $x \vee (y \wedge z) \leq (x \vee y) \wedge (x \vee z)$
 (iii) $(x \wedge y) \vee (y \wedge z) \vee (z \wedge x) \leq (x \vee y) \wedge (y \vee z) \wedge (z \vee x)$
 (iv) $(x \wedge y) \vee (x \wedge z) \leq x \wedge (y \vee (x \wedge z))$

REMARK.   (i)–(iii) are called *distributive inequalities*, and (iv) is the *modular inequality*.

PROOF.   We prove (iv) as an example and leave the rest to the reader. Since $x \wedge y \leq x, x \wedge z \leq x$, we obtain $(x \wedge y) \vee (x \wedge z) \leq x, x \wedge y \leq y \leq y \vee (x \wedge z)$, and $x \wedge z \leq y \vee (x \wedge z)$; therefore, $(x \wedge y) \vee (x \wedge z) \leq y \vee (x \wedge z)$. The meet of the two inequalities $(x \wedge y) \vee (x \wedge z) \leq x$ and $(x \wedge y) \vee (x \wedge z) \leq y \vee (x \wedge z)$ yields (iv). ●

LEMMA 10.   *Consider the following two identities and inequality:*
 (i) $(x \wedge y) \vee (x \wedge z) = x \wedge (y \vee z)$
 (ii) $(x \vee y) \wedge (x \vee z) = x \vee (y \wedge z)$
 (iii) $(x \vee y) \wedge z \leq x \vee (y \wedge z)$
Then (i), (ii), *and* (iii) *are equivalent in any lattice L.*

REMARK. A lattice satisfying identities (i) or (ii) is called *distributive*. Note that (i) and (ii) are *not* equivalent for fixed elements; that is, (i) can hold for three elements $a,b,c \in L$, whereas (ii) does not.

PROOF. Let (i) hold in L and let $a,b,c \in L$; then, using (i) with $x = a \vee b$, $y = a, z = c$,

$$(a \vee b) \wedge (a \vee c) = ((a \vee b) \wedge a) \vee ((a \vee b) \wedge c)$$

and using $a = (a \vee b) \wedge a$ and (i) for $x = c, y = a, z = b$,

$$= a \vee (a \wedge c) \vee (b \wedge c)$$
$$= a \vee (b \wedge c),$$

verifying (ii).

The proof of (ii) implies (i) is the dual of the preceding paragraph.

Let (ii) hold in L; then $x \vee (y \wedge z) = (x \vee y) \wedge (x \vee z) \geq (x \vee y) \wedge z$, since $x \vee z \geq z$, verifying (iii).

Let (iii) hold in L. Then, with $x = a, y = b, z = a \vee c$ in (iii),

$$(a \vee b) \wedge (a \vee c) \leq a \vee (b \wedge (a \vee c)) = a \vee ((a \vee c) \wedge b)$$

and with $x = a, y = c, z = c$ in (iii)

$$\leq a \vee (a \vee (c \wedge b)) = a \vee (c \wedge b).$$

This, combined with the dual of Lemma 9(i), gives (ii). ●

COROLLARY 11. *The dual of a distributive lattice is distributive.*

LEMMA 12. *The identity*

$$(x \wedge y) \vee (x \wedge z) = x \wedge (y \vee (x \wedge z))$$

is equivalent to the condition:

$$x \geq z \text{ implies that } (x \wedge y) \vee z = x \wedge (y \vee z).$$

REMARK. A lattice satisfying either condition is called *modular*.

PROOF. If $x \geq z$, then $z = x \wedge z$; thus the implication follows from the identity. Conversely, if the implication holds, then since $x \geq x \wedge z$, we have $(x \wedge y) \vee (x \wedge z) = x \wedge (y \vee (x \wedge z))$. ●

Exercises

1. Give a formal proof of Lemma 3.
2. Let $H = \{h_0, \ldots, h_{n-1}\}$. Prove that $a \in [H]$ iff $a = p(h_0, \ldots, h_{n-1})$ for some n-ary polynomial p.
3. Show that the upper bound in Corollary 4 is best possible if $|H| \geq 3$. Give the best estimates for $|H| \leq 2$.
4. Give a formal proof of Lemma 8.
5. Prove (without reference to Lemma 8) that if L is distributive (modular), then so is $I(L)$.
6. Show that the dual of a modular lattice is modular.
7. Prove that L is distributive iff the identity
$$(x \wedge y) \vee (y \wedge z) \vee (z \wedge x) = (x \vee y) \wedge (y \vee z) \wedge (z \vee x)$$
holds in L.
8. Prove that every distributive lattice is modular, but not conversely. Find the smallest modular but not distributive lattice.
9. Find an identity $p = q$ characterizing distributive lattices such that neither $p \leq q$ nor $q \leq p$ holds in a general lattice.
10. Show that in any lattice
$$\bigvee (\bigwedge (x_{ij} \mid 0 \leq i < m) \mid 0 \leq j < n)$$
$$\leq \bigwedge (\bigvee (x_{ij} \mid 0 \leq j < n) \mid 0 \leq i < m).$$
11. Prove that the following identity holds in a distributive lattice:
$$\bigvee (\bigwedge (x_{ij} \mid 0 \leq j < n) \mid 0 \leq i < m)$$
$$= \bigwedge (\bigvee (x_{if(i)} \mid 0 \leq i < m) \mid f \in F),$$
where F is the set of all functions from $\{0, 1, \ldots, m - 1\}$ into $\{0, 1, \ldots, n - 1\}$.
12. Derive (i)–(iii) of Lemma 9 from exercise 10.
13. Verify that any chain is a distributive lattice.
14. Let L be a lattice with more than one element, let $L' = L \cup \{0\}$, and let $0 < x$ for all $x \in L$. Show that L' is then a lattice and that an identity $p = q$ holds in L iff it holds in L'.
15. Show that if the identity $x_0 = p(x_0, x_1, \ldots)$ holds in the two-element lattice, then it holds in every lattice.
16. Prove that $P(X)$ is a distributive lattice.
17. The set of all equivalence relations on the set X partially ordered under set inclusion forms a lattice $E(X)$. Show that $E(X)$ is distributive iff $|X| \leq 2$ and modular iff $|X| \leq 3$.
18. Find an identity that holds in $\mathfrak{C}_2$ but not in $\mathfrak{N}_5$.
19. Find an identity that holds in $\mathfrak{C}_2$ but not in $\mathfrak{M}_5$.
20. Let K and L be modular lattices, let D be a dual ideal of K, and let I be an ideal of L such that there exists an isomorphism φ of D with I. Set $A = K \cup (L - I)$ (disjoint union) and define $x \leq y$, for $x, y \in A$ as

follows: For $x,y \in K$, or $x,y \in L - I$, $x \le y$ retains its meaning; for $x \in K$, $y \in L - I$, let $x \le y$ mean the existence of a $z \in D$ such that $x \le z$, $z\varphi \le y$; $x \le y$ for no $x \in L - I$, $y \in K$. Is $\langle A; \le \rangle$ then a modular lattice?

21. If K and L of exercise 20 are distributive, is A distributive?
22. In exercise 20, put $I = D = \{a\}$ and show that any identity holding in K and L holds in A as well.
23. Examine the statements of Lemma 8 for properties of the form, "If $p_0 = q_0$, then $p_1 = q_1$."
*24. Show that $\langle L; \wedge, \vee \rangle$ is a lattice satisfying the identity $p = q$ iff it satisfies the identities

$$(w \wedge x) \vee x = x$$

and

$$[((x \wedge p) \wedge z) \vee u] \vee v = [((q \wedge z) \wedge x) \vee v] \vee [(t \vee u) \wedge u],$$

where x, z, u, v, and w are variables that do not occur in p or q (R. Padmanabhan [1968]).
*25. Show that the result of exercise 24 is best possible: If $p = q$ is an identity *not* satisfied by some lattice, then the two identities of exercise 24 cannot be replaced by one (R. N. McKenzie [a]).
26. Show that the lattice L is modular iff the inequality $x \wedge (y \vee z) \le y \vee ((x \vee y) \wedge z)$ holds in L.

5. Free Lattices

Though it would be quite easy to develop a feeling for the most general lattice (generated by a set of elements and satisfying some relation) of Section 2 by way of some examples, a general definition seems to be hard to accept. So we ask the reader to withhold judgment on whether Definition 2 expresses his intuitive feelings until the theory is developed and further examples are presented.

The most general lattice will be called free. Since we might be interested, for instance, in the most general distributive lattice generated by a, b, c, satisfying $b < a$, it seems desirable to define freeness with respect to a class **K** of lattices.

DEFINITION 1. *Let $p_i = q_i$ be identities for $i \in I$. The class* **K** *of all lattices satisfying all identities $p_i = q_i$, $i \in I$, is called an* equational class of lattices. *An equational class is* trivial *if it contains one-element lattices only.*

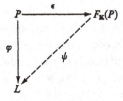

Figure 5.1

The class **L** of all lattices, the class **D** of all distributive lattices, and the class **M** of all modular lattices are examples of equational classes of lattices.

Next we have to agree on what kinds of relations to allow in the generating set. Can we prescribe only relations of the form $b < a$, or do we allow relations of the form $a \wedge b = c$ or $d \vee e = f$? Lemma 9 is an example in which the four generators x, y, z, u are required to satisfy $x \vee y = x \vee z = y \vee z = u$. Let us therefore agree that for a generating set we take a poset P, and for relations we take all $a \leq b$ that hold in P, all $a \wedge b = c$, where inf $\{a, b\} = c$ in P, and all $a \vee b = c$, where sup $\{a, b\} = c$ in P. (A more liberal approach will be presented later in this section.)

Now we are ready to formulate our basic concept:

DEFINITION 2.   *Let P be a poset and let* **K** *be an equational class of lattices. A lattice $F_{\mathbf{K}}(P)$ is called a* free lattice over **K** generated by P *if the following conditions are satisfied:*

(i) *$F_{\mathbf{K}}(P) \in$ **K**.*

(ii) *$P \subseteq F_{\mathbf{K}}(P)$, and for $a,b,c \in P$, inf $\{a, b\} = c$ in P iff $a \wedge b = c$ in $F_{\mathbf{K}}(P)$, and sup $\{a, b\} = c$ in P iff $a \vee b = c$ in $F_{\mathbf{K}}(P)$.*

(iii) *$[P] = F_{\mathbf{K}}(P)$.*

(iv) *Let $L \in$ **K** and let $\varphi : P \to L$ be an isotone map with the properties that if $a,b,c \in P$, inf $\{a, b\} = c$ in P, then $a\varphi \wedge b\varphi = c\varphi$ in L, and if sup $\{a, b\} = c$ in P, then $a\varphi \vee b\varphi = c\varphi$ in L. Then there exists a (lattice) homomorphism $\psi : F_{\mathbf{K}}(P) \to L$ extending φ (that is, $a\varphi = a\psi$ for all $a \in P$).*

Let ϵ denote the identity map on P; then the crucial condition (iv) can be expressed by Figure 5.1. In that and in all such similar *diagrams* (not in the sense the word is used in Section 2), the capital letters represent lattices or posets, and the arrows indicate homomorphisms, or maps with certain properties. Such diagrams are usually assumed to be *commutative*, which, in this particular case, means $\epsilon\psi = \varphi$, which is (iv). Note that (ii) is also included in the diagram if the arrows are supposed to represent maps as required in (iv). In fact, we could have required in (ii) that the identity map on P be an embedding of P into $F_K(P)$ in the sense of (iv).

If P is an unordered set, $|P| = \mathfrak{m}$, we shall write $F_K(\mathfrak{m})$ for $F_K(P)$ and call it a *free lattice on $\mathfrak{m}$ generators* over K. In case $K = L$, we shall always omit "over L"; thus "free lattice generated by P" shall mean "free lattice over L generated by P"—in notation, $F(P)$.

It should be noted that if $b \in F_K(P)$, then by (iii) and by Lemma 4.3, $b = p(a_0, \ldots, a_{n-1})$, where p is a polynomial and $a_0, \ldots, a_{n-1} \in P$. Thus if the ψ of (iv) exists, then we must have

$$b\psi = p(a_0, \ldots, a_{n-1})\psi \quad \text{(since } \psi \text{ is to be a homomorphism)}$$

$$= p(a_0\psi, \ldots, a_{n-1}\psi) = p(a_0\varphi, \ldots, a_{n-1}\varphi),$$

since $a_i\psi = a_i(\epsilon\psi) = a_i\varphi$.

From this we conclude:

COROLLARY 3. *The homomorphism ψ in* (iv) *is unique.*

This corollary is used to prove:

COROLLARY 4. *Let both $F_K(P)$ and $F_K^*(P)$ satisfy the conditions of Definition 2. Then there exists an isomorphism $\chi: F_K(P) \to F_K^*(P)$, and χ can be chosen so that $a\chi = a$ for all $a \in P$. In other words, free lattices (over K generated by P) are unique up to isomorphism.*

PROOF. Let us use Figure 5.1 with $L = F_K(P)$ and $\varphi = \epsilon$. Then there exist $\psi_1: F_K(P) \to F_K^*(P)$ and $\psi_2: F_K^*(P) \to F_K(P)$ such that $\epsilon\psi_1 = \epsilon$ and $\epsilon\psi_2 = \epsilon$. Thus $\psi_1\psi_2: F_K(P) \to F_K(P)$ is the identity map ϵ on P. By Corollary 3, ϵ has a unique extension to a homomorphism $F_K(P) \to F_K(P)$; the identity map on $F_K(P)$ is one such extension. Therefore, $\psi_1\psi_2$ is the identity map on $F_K(P)$. Similarly, $\psi_2\psi_1$ is the identity map on $F_K^*(P)$, and so (exercise 3.2) ψ_1 is the required isomorphism. ●

This settles the uniqueness, but how about existence? Naturally, $F_K(P)$ need not exist. For instance, $F_D(\mathfrak{N}_5)$ should be $\mathfrak{N}_5$, since by (ii) and (iii) of Definition 2, $F_K(P) = P$ if P is a lattice; but $\mathfrak{N}_5 \notin D$, so (i) is violated.

THEOREM 5. *Let P be a poset and let* **K** *be an equational class of lattices. Then $F_{\mathbf{K}}(P)$ exists iff the following condition is satisfied:*

(E)   *There exists a lattice L in* **K** *such that $P \subseteq L$ and for $a,b,c \in P$, inf $\{a, b\} = c$ in P iff $a \wedge b = c$ in L, and sup $\{a, b\} = c$ in P iff $a \vee b = c$ in L.*

PROOF.   Condition (E) is obviously necessary for the existence of $F_{\mathbf{K}}(P)$; indeed, if $F_{\mathbf{K}}(P)$ exists, (E) can always be satisfied with $L = F_{\mathbf{K}}(P)$ by (i) and (ii) of Definition 2.

Now assume that (E) is satisfied. In this proof a map $\varphi: P \to N$ ($N \in \mathbf{K}$) will be called a *homomorphism* if it satisfies the conditions set forth in Definition 2(iv).

Obviously, in Definition 2(iv) it suffices to consider N with $N = [P\varphi]$. Let $\langle N, \varphi \rangle$ denote this situation—that is, $N \in \mathbf{K}$, $\varphi: P \to N$ is a homomorphism and $N = [P\varphi]$. Then $F_{\mathbf{K}}(P)$, or, more precisely, $\langle F_{\mathbf{K}}(P), \epsilon \rangle$ has the property that for any $\langle L, \varphi \rangle$ there exists a $\psi: F_{\mathbf{K}}(P) \to L$ with $\varphi = \epsilon\psi$. To construct $F_{\mathbf{K}}(P)$ we have to construct a lattice having this property for *all* $\langle L, \varphi \rangle$. How would we construct one for only two?

Let $\langle L_1, \varphi_1 \rangle$ and $\langle L_2, \varphi_2 \rangle$ be given. Form $L_1 \times L_2$ and define a map $\varphi: P \to L_1 \times L_2$ by $p\varphi = \langle p\varphi_1, p\varphi_2 \rangle$; set $N = [P\varphi]$. The fact that φ is a homomorphism is easy to check. A simple example is illustrated in Figures 5.2–5.4. Now we define $\psi_i: \langle x_1, x_2 \rangle \to x_i$ and obviously, for $p \in P$, $p\varphi\psi_i = p\varphi_i$ and $\psi_i: N \to L_i$.

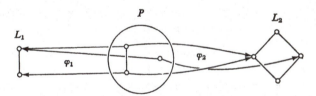

Figure 5.2

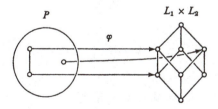

Figure 5.3

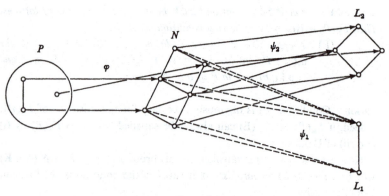

Figure 5.4

If we are given any number of $\langle L_i, \varphi_i \rangle$, $i \in I$, we can proceed as before and get $\langle N, \varphi \rangle$; if one of the $\langle L_i, \varphi_i \rangle$ is the $\langle L, \epsilon \rangle$ given by (E), then (ii) of Definition 2 will also be satisfied. There is only one problem: All the pairs $\langle L_i, \varphi_i \rangle$ do not form a set, so their direct product cannot be formed. The $\langle L_i, \varphi_i \rangle$ do not form a set because a lattice and all its isomorphic copies do not form a set; therefore, if we can somehow restrict taking too many isomorphic copies, the previous procedure can be followed. Observe that by Corollary 4.4, in every pair $\langle L_i, \varphi_i \rangle$ we have

$$|L_i| \le |P\varphi_i| + \aleph_0 \le |P| + \aleph_0.$$

Thus, by choosing a large enough set S and taking only those $\langle L_i, \varphi_i \rangle$ that satisfy $L_i \subseteq S$, we can solve our problem.

Now we are ready to proceed with the formal proof. Choose a set S satisfying $|P| + \aleph_0 = |S|$. Let Q be the set of all pairs $\langle M, \psi \rangle$ where $M \subseteq S$. Form $A = \prod (M \mid \langle M, \psi \rangle \in Q)$, and for each $p \in P$ let $f_p \in A$ be defined by

$$f_p(\langle M, \psi \rangle) = p\psi.$$

Finally, set

$$N = [\{f_p \mid p \in P\}].$$

We claim that if, for all $p \in P$, we identify p with f_p, then N satisfies (i)–(iv) of Definition 2, and thus $N = F_K(P)$.

(i) N is constructed from members of K by forming a direct product and by taking a sublattice. By Lemma 4.8, $N \in \mathbf{K}$, since $\mathbf{K}$ is an equational class.

(ii) Let inf $\{a, b\} = c$ in P. Then for every $\langle M, \psi \rangle \in Q$, $a\psi \wedge b\psi = c\psi$, so that $f_a(\langle M, \psi \rangle) \wedge f_b(\langle M, \psi \rangle) = f_c(\langle M, \psi \rangle)$, that is, $f_a \wedge f_b = f_c$. Since p is identified with f_p, we conclude that $a \wedge b = c$ in N.

   Conversely, let $a \wedge b = c$ in N, that is, $f_a \wedge f_b = f_c$. Let L be the lattice given by (E) and let ϵ be the identity map on P; then we can form $\langle L, \epsilon \rangle$. By Corollary 4.4, $|L| \leq |S|$, so there is a one-to-one map $\alpha : L \rightarrow S$. Let $L_1 = L\alpha$ and make L_1 into a lattice by defining

$$a\alpha \wedge b\alpha = (a \wedge b)\alpha, \qquad a\alpha \vee b\alpha = (a \vee b)\alpha.$$

Then $L \cong L_1$ and we can form the pair $\langle L_1, \alpha_1 \rangle$, where α_1 is the restriction of α to $P (\subseteq L)$. Since $L_1 \subseteq S$, $\langle L_1, \alpha_1 \rangle \in Q$. Now $f_a \wedge f_b = f_c$ yields

$$f_a(\langle L_1, \alpha_1 \rangle) \wedge f_b(\langle L_1, \alpha_1 \rangle) = f_c(\langle L_1, \alpha_1 \rangle);$$

that is, $a\alpha \wedge b\alpha = c\alpha$, which in turn gives $a \wedge b = c$, since α is an isomorphism. By (E), $a \wedge b = c$ in L implies that inf $\{a, b\} = c$. The second part of (ii) follows by duality.

(iii) This part of the proof is obvious by the definition of N.

(iv) Take $\langle L, \varphi \rangle$; we have to find a homomorphism $\psi : N \rightarrow L$ satisfying $a\varphi = a\psi$ for $a \in P$. Using $|N| \leq |S|$, the argument given in (ii) can be repeated to find $\langle L_1, \varphi_1 \rangle$, an isomorphism $\alpha : L \rightarrow L_1$ such that $a\varphi\alpha = a\varphi_1$ for all $a \in P$ and $L_1 \subseteq S$. Therefore, $\langle L_1, \varphi_1 \rangle \in Q$. Set

$$\psi_1 : f \rightarrow f(\langle L_1, \varphi_1 \rangle), \qquad f \in N.$$

Then for $a \in P$,

$$a\psi_1 = f_a\psi_1 = f_a(\langle L_1, \varphi_1 \rangle) = a\varphi_1 = a\varphi\alpha.$$

Thus the homomorphism $\psi = \psi_1\alpha^{-1} : N \rightarrow L$ will satisfy the requirement of (iv). ●

Two consequences of this theorem are very important:

COROLLARY 6.   *For any nontrivial equational class* $\mathbf{K}$ *and for any cardinal* $\mathfrak{m}$, *a free lattice over* $\mathbf{K}$ *with* $\mathfrak{m}$ *generators*, $F_{\mathbf{K}}(\mathfrak{m})$, *exists.*

PROOF.   It suffices to find an $L \in \mathbf{K}$, $X \subseteq L$ such that $|X| = \mathfrak{m}$, and for $x, y \in X$, $x \neq y$, x and y are incomparable. This is easily done. Since $\mathbf{K}$ is nontrivial, there exists an $N \in \mathbf{K}$, $|N| > 1$; thus, $\mathfrak{C}_2$ is a sublattice of N. By Lemma 4.8, $\mathfrak{C}_2 \in \mathbf{K}$; by Lemma 4.8, $(\mathfrak{C}_2)^I \in \mathbf{K}$ for any set I. Let $|I| = \mathfrak{m}$, let $L = (\mathfrak{C}_2)^I$; for $i \in I$, define $f_i \in L$ by $f_i(i) = 1$, $f_i(j) = 0$ for $i \neq j$, and set $X = \{f_i \mid i \in I\}$. Obviously, X satisfies the condition. ●

COROLLARY 7.   *For any poset P, a free lattice (over $\mathbf{L}$) generated by the poset P exists.*

PROOF.   Take a poset P and define $I_0(P)$ to be the set of all subsets I of P satisfying the condition: sup $\{a, b\} \in I$ iff $a,b \in I$. Partially ordering $I_0(P)$ by set inclusion makes $I_0(P)$ a lattice. Identifying a with $\{x \mid x \leq a\}$, we see that $I_0(P)$ contains P and satisfies (E) of Theorem 5. The detailed computation is almost the same as that for Theorem 20, so it will be omitted. ●

The argument given in the proof of Corollary 3 shows that whenever L is generated by P, any homomorphism φ of P has at most one extension to L, and if there is one, it is given by

$$\psi: p(a_0, \ldots, a_{n-1}) \to p(a_0\varphi, \ldots, a_{n-1}\varphi).$$

This formula gives a homomorphism iff ψ is well defined; in other words, iff $p(a_0, \ldots, a_{n-1}) = q(b_0, \ldots, b_{m-1})$ implies that $p(a_0\varphi, \ldots, a_{n-1}\varphi) = q(b_0\varphi, \ldots, b_{m-1}\varphi)$, for any $a_0, \ldots, a_{n-1}, b_0, \ldots, b_{m-1} \in P$ and $\varphi: P \to N \in \mathbf{K}$.

This yields a very practical method of finding free lattices and verifying their freeness.

THEOREM 8.   *In the definition of $F_\mathbf{K}(P)$, (iv) can be replaced by the following condition:*

> *If $b \in F_\mathbf{K}(P)$ has two representations, $b = p(a_0, \ldots, a_{n-1})$ and $b = q(b_0, \ldots, b_{m-1})$ $(a_0, \ldots, a_{n-1}, b_0, \ldots, b_{m-1} \in P)$, then*
>
> $$p(a_0, \ldots, a_{n-1}) = q(b_0, \ldots, b_{m-1})$$
>
> *can be derived from the identities defining $\mathbf{K}$ and the relations of P of the form $a \wedge b = c$ and $a \vee b = c$.*

REMARK.   Thus, in proving $p = q$, we can use only the $\wedge$ and $\vee$ table of P, but we cannot use $a \neq b$ or $a \neq b \vee c$, and so on.

We illustrate Theorem 8 first by determining $F_\mathbf{D}(3)$. The following simple observation will be useful:

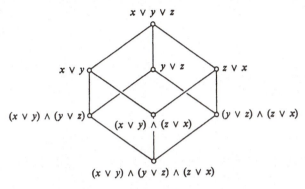

Figure 5.5

LEMMA 9.   *Let x, y, and z be elements of a lattice L and let $x \vee y$, $y \vee z$, $z \vee x$ be pairwise incomparable. Then $\{x \vee y, y \vee z, z \vee x\}$ generates a sublattice of L isomorphic to $(\mathfrak{C}_2)^3$ (see Figure 5.5).*

PROOF.   Almost all the meets and joins are obvious; by symmetry, the nonobvious ones are typified by $[(x \vee y) \wedge (y \vee z)] \vee [(x \vee y) \wedge (z \vee x)] = x \vee y$ and $[(x \vee y) \wedge (y \vee z)] \vee (z \vee x) = x \vee y \vee z$. Since $y \leq (x \vee y) \wedge (y \vee z)$ and $x \leq (x \vee y) \wedge (z \vee x)$, we get $x \vee y \leq [(x \vee y) \wedge (y \vee z)] \vee [(x \vee y) \wedge (z \vee x)]$, and $\geq$ is trivial. The second equality follows from $y \leq (x \vee y) \wedge (y \vee z)$. Note that, for example, $(x \vee y) \wedge (y \vee z) = (x \vee y) \wedge (y \vee z) \wedge (z \vee x)$ would imply, by joining both sides with $y \vee z$, that $x \vee y \vee z = z \vee x$; thus $x \vee y \leq z \vee x$, a contradiction. Therefore, all eight elements are distinct.   ●

THEOREM 10.   *A free distributive lattice on three generators $F_D(3)$ has eighteen elements (see Figure 5.6).*

PROOF.   Let x, y, and z be the free generators. The top eight and the bottom eight elements form sublattices by Lemma 9 and its dual; note that $(x \wedge y) \vee (y \wedge z) \vee (z \wedge x) = (x \vee y) \wedge (y \vee z) \wedge (z \vee x)$ by exercise 4.7.

According to Theorem 8, we have only to verify that the lattice L of Figure 5.6 is a distributive lattice, and that if p, q, r are polynomials representing elements of L and $p \wedge q = r$ in L, then $p \wedge q = r$ in every distributive lattice and similarly for $\vee$. The first statement is easily proved by

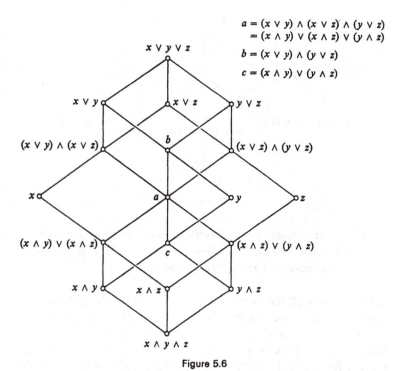

Figure 5.6

representing L by sets (see exercise 13). The second statement requires a complete listing of all triples p, q, r with $p \wedge q = r$. If p, q, r belong to the top eight or bottom eight elements, the statement follows from Lemma 9. The remaining cases are all trivial except when p or q is x, y, or z. By symmetry, only $p = x$, $q = y \vee z$, $r = (x \wedge y) \vee (x \wedge z)$ is left to discuss, but then $p \wedge q = r$ is the distributive law. ●

THEOREM 11   (R. Dedekind [1900]).   *A free modular lattice on three generators $F_M(3)$ has twenty-eight elements (see Figure 5.7).*

PROOF.   Let x, y, and z be the free generators. Again modularity is proved by a representation (see exercise 16). Theorem 10 takes care of most meets and joins not involving x_1, y_1, z_1. Of the rest, only one relation (and the symmetric and the dual cases) is difficult to prove: $x_1 \wedge y_1 = u$. This we do now, leaving the rest to the reader.

$u = (x \wedge y) \vee (y \wedge z) \vee (x \wedge z)$

$v = (x \vee y) \wedge (y \vee z) \wedge (x \vee z)$

$x_1 = (x \wedge v) \vee u$

$y_1 = (y \wedge v) \vee u$

$z_1 = (z \wedge v) \vee u$

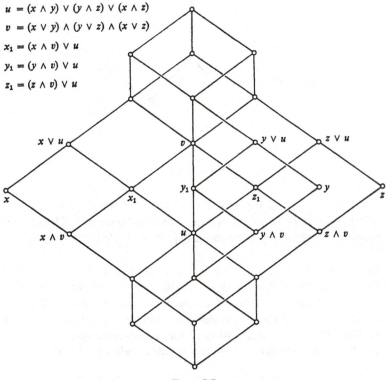

Figure 5.7

Compute:

$$x_1 \wedge y_1 = [(x \wedge v) \vee u] \wedge [(y \wedge v) \vee u] \quad (\text{since } u \leq (y \wedge v) \vee u)$$

$$= [(x \wedge v) \wedge ((y \wedge v) \vee u)] \vee u$$

$$= [(x \wedge v) \wedge (y \vee u) \wedge v] \vee u \quad (\text{substitute } u \text{ and } v)$$

$$= [x \wedge (y \vee z) \wedge (y \vee (x \wedge z))] \vee u$$

$$= (x \wedge y) \vee (x \wedge z) \vee u = u. \quad \bullet$$

Consider the lattice represented by Figure 5.8. We would like to say that it is freely generated by $\{0, a, b, 1\} = P$, but this is clearly not the case according to Definition 2, since sup $\{a, b\} = 1$ in P, whereas in the lattice,

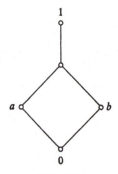

Figure 5.8

$a \vee b < 1$. So to get the most general lattice of Section 2 we have to enlarge the framework of our discussion by introducing partial lattices. Of course, the study of partial lattices is important also for other purposes.

DEFINITION 12.   *Let L be a lattice, $H \subseteq L$, and restrict $\wedge$ and $\vee$ to H as follows: For $a,b,c \in H$, if $a \wedge b = c$ (dually, $a \vee b = c$), then we say that in H $a \wedge b$ (dually, $a \vee b$) is defined and it equals c; if, for $a,b \in H$, $a \wedge b$ (dually, $a \vee b$) $\notin H$, then we say that $a \wedge b$ (dually, $a \vee b$) is not defined in H. Thus $\langle H; \wedge, \vee \rangle$ is a set with two binary partial operations. $\langle H; \wedge, \vee \rangle$ is called a partial lattice. $\langle H; \wedge, \vee \rangle$ is called a relative sublattice of L.*

Thus every subset of a lattice determines a partial lattice. The second part of this section is devoted to an internal characterization of partial lattices, based on N. Funayama [1953].

We now analyze the way the eight identities that were used to define lattices ((L1)–(L4) of Section 1) hold in partial lattices:

LEMMA 13.   *Let $\langle H; \wedge, \vee \rangle$ be a partial lattice, $a,b,c \in H$.*

(i) *$a \wedge a$ exists, and $a \wedge a = a$.*

(ii) *If $a \wedge b$ exists, then $b \wedge a$ exists, and $a \wedge b = b \wedge a$.*

(iii) *If $a \wedge b$, $(a \wedge b) \wedge c$, $b \wedge c$ exist, then $a \wedge (b \wedge c)$ exists, and $(a \wedge b) \wedge c = a \wedge (b \wedge c)$. If $b \wedge c$, $a \wedge (b \wedge c)$, $a \wedge b$ exist, then $(a \wedge b) \wedge c$ exists, and $(a \wedge b) \wedge c = a \wedge (b \wedge c)$.*

(iv) *If $a \wedge b$ exists, then $a \vee (a \wedge b)$ exists, and $a = a \vee (a \wedge b)$.*

PROOF.   As an illustration let us prove the first statement of (iii). Let $H \subseteq L$ as in Definition 12. The assumption that $a \wedge b$, $(a \wedge b) \wedge c$,

$b \wedge c$ exist in H means that in L, $a, b, c, a \wedge b, (a \wedge b) \wedge c, b \wedge c \in H$. But in L, $(a \wedge b) \wedge c = a \wedge (b \wedge c)$, and so $a \wedge (b \wedge c) \in H$; that is, $a \wedge (b \wedge c)$ exists in H, and, of course, $(a \wedge b) \wedge c = a \wedge (b \wedge c)$. ●

LEMMA 13'. *Let* (i')–(iv') *denote the statements we get from* (i)–(iv) *of Lemma 13 by interchanging* $\wedge$ *and* $\vee$ *. Then* (i')–(iv') *hold in any partial lattice.*

PROOF. This proof is trivial by duality. ●

Statements (i)–(iv) and (i')–(iv') give the required interpretation of the eight identities for partial lattices.

DEFINITION 14. *A* weak partial lattice *is a set with two binary partial operations satisfying* (i)–(iv) *and* (i')–(iv').

COROLLARY 15. *Every partial lattice is a weak partial lattice.*

Based on Figure 5.9, we give the following example of a weak partial lattice that is not a partial lattice.

Let $H = \{0, a, b, c, d, e, f, g, h, 1\}$. Consider the lattice $\mathfrak{H}$ of Figure 5.9. Define $x \wedge y = z$ in H if $x \wedge y = z$ in $\mathfrak{H}$. Define $x \vee y = z$ in H if either $x \le y$ in $\mathfrak{H}$ and $y = z$, or $y \le x$ in $\mathfrak{H}$ and $x = z$; or if $\{x, y\} = \{a, c\}$, and $z = f$; or $\{x, y\} = \{b, d\}$, and $z = g$; or $\{x, y\} = \{f, g\}$, and $z = 1$. Then $\langle H; \wedge, \vee \rangle$ is a weak partial lattice (check the axioms). Now suppose that there exists a lattice L, $H \subseteq L$, such that $\langle H; \wedge, \vee \rangle$ is a relative sublattice

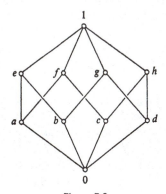

Figure 5.9

of L. Then $1 = (a \vee c) \vee (b \vee d)$ in L, and thus $1 = \sup \{a, b, c, d\}$. Since $e \geq a,b$, and $h \geq c,d$ in L, and $1 \geq e,h$, we get $1 = \sup \{e, h\}$ in L. The fact that $e,h,1 \in H$ implies that $e \vee h$ is defined in H (and equals 1), contrary to the definition of $\langle H; \wedge, \vee \rangle$ (compare this with Lemma 19).

To avoid such anomalies we shall introduce two further conditions. To prepare for them we prove:

LEMMA 16.   *Let* $\langle H; \wedge, \vee \rangle$ *be a weak partial lattice. Then we define a partial ordering relation* $\leq$ *on* H *by* "$a \leq b$ *iff* $a \wedge b$ *exists and* $a \wedge b = a$." *If* $a \vee b$ *exists, then* $a \vee b = \sup \{a, b\}$. *If* $a \wedge b$ *exists, then* $a \wedge b = \inf \{a, b\}$. *Also,* $a \leq b$ *iff* $a \vee b = b$.

PROOF.   This proof is the same as the proof of the corresponding parts of Theorem 1.1, except that the arguments are a bit longer.   ●

Note that in a partial lattice sup $\{a, b\}$ may exist but $a \vee b$ does not. For instance, let L be the lattice of Figure 5.9, $H = \{0, a, b, 1\}$. Then sup $\{a, b\}$ = 1 in H, but $a \vee b$ is not defined in H because $a \vee b \notin H$.

DEFINITION 17.   *An* ideal *of a weak partial lattice* H *is a nonvoid subset* I *of* H *such that if* $a,b \in I$ *and* $a \vee b$ *exists, then* $a \vee b \in I$, *and* $x \leq a \in I$ *implies that* $x \in I$. *Again we set* $\{x \mid x \leq a\} = (a]$. Dual ideal *and* $[a)$ *are defined dually.* $I_0(H)$ *is the lattice consisting of* $\varnothing$ *and all ideals of* H *(partial ordering is* $\subseteq$), $\mathcal{D}_0(H)$ *is the lattice consisting of* $\varnothing$ *and all dual ideals of* H *(partial ordering is* $\subseteq$). *For* $K \subseteq H$, $(K]$ *is the ideal and* $[K)$ *is the dual ideal generated by* K.

COROLLARY 18.   *Let* H *and* L *be given as in Definition 12. Let* I *be an ideal of* L. *Then* $I \cap H$ *is an ideal of* H *provided that* $I \cap H \neq \varnothing$.

LEMMA 19.   *Any partial lattice satisfies the following condition:*
   (I) *If* $(a] \vee (b] = (c]$ *in* $I_0(H)$, *then* $a \vee b$ *exists in* H *and equals* c.

PROOF.   Let H and L be given as in Definition 12, let $a,b,c \in H$, and let $(a] \vee (b] = (c]$ in $I_0(H)$. Set $I = (a \vee b]_L$. Then $(a]_H, (b]_H \subseteq I \cap H$, thus $(c]_H = (a]_H \vee (b]_H \subseteq (a \vee b]_L$, that is, $c \leq a \vee b$. Since $a \leq c, b \leq c$, we conclude that $a \vee b = c$.   ●

Let (D) denote the condition dual to (I), namely:
   (D)   If $[a) \vee [b) = [c)$ in $\mathcal{D}_0(H)$, then $a \wedge b$ exists in H and equals c.

THEOREM 20   (N. Funayama [1953]).   *A partial lattice is a weak partial lattice satisfying conditions* (I) *and* (D).

PROOF.   Corollary 14 and Lemma 19 and its dual prove that a partial lattice is a weak partial lattice satisfying (I) and (D). Conversely, let $\langle H; \wedge, \vee \rangle$ be a weak partial lattice satisfying (I) and (D). Consider the map

$$\varphi: x \to \langle (x], [x) \rangle,$$

sending H into $I_0(H) \times \breve{\mathscr{D}}_0(H)$, where $\breve{\mathscr{D}}_0(H)$ is the dual of $\mathscr{D}_0(H)$. This map φ is one-to-one. If $x \vee y = z$, then $(x] \vee (y] = (z]$ in $I_0(H)$ and $[x) \vee [y) = [z)$ in $\breve{\mathscr{D}}_0(H)$, thus $x\varphi \vee y\varphi = (x \vee y)\varphi$. Conversely, if $x\varphi \vee y\varphi = z\varphi$, then $(x] \vee (y] = (z]$ in $I_0(H)$. Therefore, by (I), $x \vee y$ exists and equals z, so $x\varphi \vee y\varphi = z\varphi$ implies that $x \vee y = z$. A similar argument shows that $x \wedge y = z$ iff $x\varphi \wedge y\varphi = z\varphi$. Thus we can identify x with $x\varphi$, getting $H \subseteq L = I_0(H) \times \breve{\mathscr{D}}_0(H)$. We have just proved that $\langle H; \wedge, \vee \rangle$ is a relative sublattice of L.   ●

Let $\langle P; \leq \rangle$ be a poset. We make P into a partial lattice as follows: $a \wedge b$ is defined iff inf $\{a, b\}$ exists and $a \wedge b = \inf \{a, b\}$, and similarly for $a \vee b$.

LEMMA 21. $\langle P; \wedge, \vee \rangle$ *is a partial lattice.*

PROOF.   It is easy to verify that (I), (D), (i)–(iv) of Lemma 13, and (i')–(iv') of Lemma 13' hold. Another proof identifies $x \in P$ with $(x]$, and thus we get $P \subseteq I(P)$. Then we observe that $\wedge$ and $\vee$ of $I(P)$ restricted to P give the $\wedge$ and $\vee$ of P.   ●

We need some further definitions.

DEFINITION 22.   *Let* $\langle A; \wedge, \vee \rangle$, $\langle B; \wedge, \vee \rangle$ *be weak partial lattices,* $\varphi: A \to B$. *We call* φ *a* homomorphism *if, whenever* $a \wedge b$ *exists for* $a, b \in A$, *then* $a\varphi \wedge b\varphi$ *exists and* $(a \wedge b)\varphi = a\varphi \wedge b\varphi$, *and if the similar condition holds for* $\vee$. *A one-to-one homomorphism* φ *is an* embedding *provided that* $a \wedge b$ *exists iff* $a\varphi \wedge b\varphi$ *exists, and if the similar condition holds for* $\vee$. *If* φ *is onto and* φ *is an embedding, then* φ *is an* isomorphism.

Now we are ready again to define the most general lattices of Section 2.

DEFINITION 23. *Let* $\mathfrak{A} = \langle A; \wedge, \vee \rangle$ *be a partial lattice and let* **K** *be an equational class of lattices. The lattice* $F_{\mathbf{K}}(\mathfrak{A})$ *(or, simply,* $F_{\mathbf{K}}(A)$*) is a* free *lattice over* **K** *generated by* $\mathfrak{A}$*, if the following conditions are satisfied:*

(i) $F_{\mathbf{K}}(A) \in \mathbf{K}$.

(ii) $A \subseteq F_{\mathbf{K}}(A)$*, and* A *is partial sublattice of* $F_{\mathbf{K}}(A)$*.*

(iii) $[A] = F_{\mathbf{K}}(A)$*.*

(iv) *If* $L \in \mathbf{K}$ *and* $\varphi: A \rightarrow L$ *is a homomorphism, then there exists a homomorphism* $\psi: F_{\mathbf{K}}(A) \rightarrow L$ *extending* φ *(that is,* $a\varphi = a\psi$ *for* $a \in A$*).*

If P is a poset, then $F_{\mathbf{K}}(P)$ is a free lattice over **K** generated by P, where P is considered a partial lattice as in Lemma 21. This theory can be developed exactly as it was in the first part of this section. The final result is:

THEOREM 24. *Let* $\mathfrak{A} = \langle A; \wedge, \vee \rangle$ *be a partial lattice and let* **K** *be an equational class. Then* $F_{\mathbf{K}}(\mathfrak{A})$ *exists iff there exists a lattice* L *in* **K** *such that* $\mathfrak{A}$ *is a relative sublattice of* L*.*

As an application we prove the existence of the lattice absolutely freely generated by a poset. Let P be a poset; we define a partial lattice P^m on P as follows:

$$x \wedge y = z \text{ in } P^m \text{ iff } x \text{ and } y \text{ are comparable and } z = \inf\{x, y\};$$

$$x \vee y = z \text{ in } P^m \text{ iff } x \text{ and } y \text{ are comparable and } z = \sup\{x, y\}.$$

DEFINITION 25. $F(P^m)$ $(= F_{\mathbf{L}}(P^m))$ *is called a lattice absolutely freely generated by* P*.*

THEOREM 26. *For any poset* P*, a lattice absolutely freely generated by* P *exists.*

PROOF. A subset $A \subseteq P$ is called *hereditary* if $x \in A$ and $y \leq x$ imply that $y \in A$. Let $H(P)$ be the set of all hereditary subsets of P partially ordered under set inclusion. Let $P_1 = H(P)$, $P_2 = H(\tilde{P}_1)$, where $\tilde{P}_1$ is the dual of P_1. Identifying $p \in P$ with $(p]$, we get $P \subseteq P_1$; identifying $p \in P_1$ with $[p)$, we get $P_1 \subseteq P_2$, thus $P \subseteq P_2$. Obviously, P_2 is a lattice. Let $a,b,c \in P$, $a \vee b = c$ in P_2. If a and b are incomparable, then $(a] \cup (b]$ $(\neq (c])$ is an upper bound for a and b, thus $a \vee b < c$ in P_2. A similar argument works for $a \wedge b = c$. Therefore, $P_2 \cong P$ satisfies the condition of Theorem 24. ●

Exercises

1. Show that an equational class is closed under the formation of sublattices, homomorphic images, and direct products.

2. Show that the class **T** of all one-element lattices is an equational class. For any equational class **K**, $\mathbf{K} \supseteq \mathbf{T}$.

3. Let $\mathbf{K}_i$, $i \in I$, be equational classes. Show that $\bigcap (\mathbf{K}_i \mid i \in I)$ is again an equational class.

4. Find an equational class **A** with $\mathbf{A} \subseteq \mathbf{M}$, $\mathbf{D} \subseteq \mathbf{A}$, and $\mathbf{A} \neq \mathbf{M}$, $\mathbf{A} \neq \mathbf{D}$. (Hint: Let **A** be defined by

$$(x \vee y) \wedge (x \vee z) \wedge (x \vee u)$$
$$= x \vee [(x \vee y) \wedge z \wedge u] \vee [(x \vee z) \wedge y \wedge u] \vee [(x \vee u) \wedge y \wedge z].)$$

5. Let **K** be a nontrivial equational class. Show that **K** contains arbitrarily large lattices.

6. Let $P = \{0, a, b\}$, $\inf \{a, b\} = 0$, and let **K** be a nontrivial equational class. Show that $F_{\mathbf{K}}(P) \cong F(2)$.

7. Find an example of an isomorphism $\varphi \colon F_{\mathbf{K}}(P) \to F_{\mathbf{K}}(P)$ that is *not* the identity map on P.

8. Let P be a poset, let **K**, **N** be equational classes, and assume that $F_{\mathbf{K}}(P)$ and $F_{\mathbf{N}}(P)$ exist. Prove that if $\mathbf{K} \supseteq \mathbf{N}$, then there exists a homomorphism φ from $F_{\mathbf{K}}(P)$ onto $F_{\mathbf{N}}(P)$ such that φ is the identity map on P.

9. Work out a proof of the existence of $F(3)$ without any reference to Theorem 5.

10. Prove Theorem 8.

11. Formulate and prove the form of Theorem 8 that is used in the proofs of Theorems 10 and 11.

12. Prove that $F_{\mathbf{M}}(4)$ is infinite (G. Birkhoff [1933]). (Hint: Let R be the set of real numbers and let L be the lattice of vector subspaces of R^3. Set

$$a = \{\langle x, 0, x \rangle \mid x \in R\}, \qquad b = \{\langle 0, x, x \rangle \mid x \in R\},$$
$$c = \{\langle 0, 0, x \rangle \mid x \in R\}, \qquad d = \{\langle x, x, x \rangle \mid x \in R\}.$$

Then $[\{a, b, c, d\}]$ is infinite.)

13. Let A, B be disjoint three-element sets. Let L be the set of the following subsets of $A \cup B$: all $X \subseteq A$, all $A \cup Y$, $Y \subseteq B$, all three-element sets Z with $|Z \cap A| = 2$. Prove that $\langle L; \subseteq \rangle$ is a lattice, that $\wedge$ and $\vee$ are intersection and union, and that thus L is distributive. Figure 5.6 is the diagram of L (A. D. Campbell [1943]).

14. Represent the lattice of Figure 5.6 as a sublattice of $(\mathfrak{C}_2)^6$.
15. Prove that six is best possible in exercise 14.
16. Represent the lattice of Figure 5.7 as a sublattice of $L \times \mathfrak{M}_5$, where L is the lattice of exercise 13 and $\mathfrak{M}_5$ is the lattice of Figure 5.10.

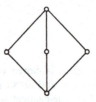

Figure 5.10

17. Show that condition (E) of Theorem 5 is equivalent to the following: For any $a,b,c \in P$ not satisfying inf $\{a, b\} = c$, there exist a lattice L in **K** and a homomorphism $\varphi: P \to L$ with $a\varphi \wedge b\varphi \neq c\varphi$, and dually.

18. The statement "$\mathfrak{A} = \langle A; \wedge, \vee \rangle$ is a *partial algebra*" means that A is a nonvoid set and that $\wedge$, $\vee$ are partial binary operations on A. For an n-ary polynomial p, $a_0, \ldots, a_{n-1} \in A$, interpret $p(a_0, \ldots, a_{n-1})$. (When is it defined and what is its value?)

19. In the weak partial lattice represented by Figure 5.9,

$$p = x \wedge [(x \vee y) \vee (x \vee z)]$$

is not defined for $x = 0$, $y = e$, $z = h$. Verify that in every lattice $p = q$, where $q = x$ and q is defined for $x = 0$.

20. An identity $p = q$ holds in the partial algebra $\mathfrak{A} = \langle A; \wedge, \vee \rangle$ if the following three conditions are satisfied:

   (i) If $p(a_0, \ldots)$, $q(a_0, \ldots)$ are defined ($a_0, \ldots \in A$), then $p(a_0, \ldots) = q(a_0, \ldots)$.

   (ii) If $p(a_0, \ldots)$ is defined, if $q = q_0 * q_1$, where $*$ is $\wedge$ or $\vee$, and if $q_0(a_0, \ldots)$, $q_1(a_0, \ldots)$ are defined, then $q_0(a_0, \ldots) * q_1(a_0, \ldots)$ is defined.

   (iii) This condition is the same as (ii) with p and q interchanged.

   Check that Lemmas 13 and 13′ give this interpretation to the lattice axioms.

21. Let $p = (((x \vee z) \vee (y \vee u)) \vee v) \vee w$ and let

$$q = [(v \vee x) \vee (v \vee y)] \vee [(w \vee z) \vee (w \vee u)].$$

   Show that $p = q$ in any lattice. Show that $p = q$ does not hold in the weak partial lattice defined in connection with Figure 5.9.

22. Let I_0, I_1 be ideals of a weak partial lattice. Set $J_0 = I_0 \cup I_1$, $J_n = \{x \mid x \leq y \vee z, y, z \in J_{n-1}\}$, $n = 1, 2, \ldots$, and $J = \bigcup (J_i \mid i = 0, 1, 2, \ldots)$. Show that $J = I_0 \vee I_1$.

23. Let the weak partial lattice L violate (I); that is, $(a] \vee (b] = (c]$, but $a \vee b$ is undefined. Let $I_0 = (a]$, $I_1 = (b]$, and $c \in J_n$ (see exercise 22).

Generalizing exercise 21, find an identity $p = q$ that holds in any lattice but not in L. (Exercise 21 is the special case in which $n = 2$.)

24. Prove that a partial algebra $\mathfrak{A} = \langle A; \wedge, \vee \rangle$ is a partial lattice iff every identity $p = q$ holding in any lattice also holds in $\mathfrak{A}$.

25. A *homomorphism* of a partial algebra $\mathfrak{A} = \langle A; \wedge, \vee \rangle$ into a lattice L is a map $\varphi: A \to L$ such that $(a \wedge b)\varphi = a\varphi \wedge b\varphi$ whenever $a \wedge b$ exists, and the same for $\vee$. Prove that there exists a one-to-one homomorphism of $\mathfrak{A}$ into some lattice L iff there exists a partial ordering $\leq$ on A satisfying $a \wedge b = \inf \{a, b\}$, whenever $a \wedge b$ is defined in $\mathfrak{A}$, and $a \vee b = \sup \{a, b\}$, whenever $a \vee b$ is defined in $\mathfrak{A}$.

26. Show that every weak partial lattice satisfies the condition of exercise 25, but not conversely.

27. Let $A = \{0, a, b, c, 1\}$, $0 \leq a, b, c \leq 1$. For $x \leq y$ define
$$x \wedge y = y \wedge x = x, \qquad x \vee y = y \vee x = y,$$
and define
$$a \wedge b = b \wedge a = 0, \qquad a \vee b = b \vee a = 1.$$
Show that $\langle A; \wedge, \vee \rangle$ is a partial lattice.

28. Let $A = \{0, a, b, c, d, 1\}$, $0 \leq a, b, c, d \leq 1$. For $x \leq y$ define
$$x \wedge y = y \wedge x = x, x \vee y = y \vee x = y,$$
and define
$$a \wedge b = b \wedge a = c \wedge d = d \wedge c = 0,$$
$$a \vee b = b \vee a = c \vee d = d \vee c = 1.$$
Show that $\langle A; \wedge, \vee \rangle$ is a partial lattice.

29. Let L and K be lattices and let $L \cap K$ be a sublattice of L and of K. For $x, y \in L \cup K$ define $x \wedge y = z$ iff $x, y, z \in L$ and $x \wedge y = z$ in L, or if $x, y, z \in K$ and $x \wedge y = z$ in K; define $x \vee y$ similarly. Is $\langle L \cup K, \wedge, \vee \rangle$ a partial lattice?

30. Are the eight axioms of a weak partial lattice independent?

31. Are the ten axioms of a partial lattice independent?

32. Define weak partial semilattice and partial semilattice; prove the analogue of Theorem 20 for partial semilattices.

33. Let A be a weak partial lattice in which $a \wedge b$ exists for all $a, b \in A$. Then A is a partial lattice iff $a \to (a]$ is an embedding of A into $I_0(A)$.

34. Let $\mathbf{T}$ be the equational class of all one-element lattices. Show that $F_{\mathbf{T}}(A)$ exists iff $|A| = 1$.

35. Let $P = (\mathbb{C}_2)^2$. Show that for any nontrivial equational class $\mathbf{K}$, $F_{\mathbf{K}}(P)$ and $F_{\mathbf{K}}(P^m)$ exist and that always $F_{\mathbf{K}}(P) \not\cong F_{\mathbf{K}}(P^m)$.

36. Determine $F(P)$, $F_{\mathbf{M}}(P)$, and $F_{\mathbf{D}}(P)$, where $P = \{a, b, c\}$ and $a < b$.

37. Discuss the set of all weak partial lattices on A inducing a given partial ordering on A.

38. Repeat exercise 37 for partial lattices.

39. Show that a one-to-one homomorphism of weak partial lattices need not be an embedding.

40. Show that in Theorem 24, the condition "there exists a lattice L in $\mathbf{K}$ such that $\mathfrak{A}$ is a relative sublattice of L" can be replaced by the following condition:

    For all $a,b,c \in A$ for which $a \wedge b = c$ does not hold, there exists a lattice L in $\mathbf{K}$, and a homomorphism φ of A into L such that $a\varphi \wedge b\varphi \neq c\varphi$, and the same condition holds for $\vee$.

41. Let $\langle A; \wedge, \vee \rangle$ be a partial algebra, let $\mathbf{K}$ be an equational class of lattices, let $L \in \mathbf{K}$, and let M be a relative sublattice of L. Then M is called a *maximal homomorphic image of A in* $\mathbf{K}$ if there is a homomorphism φ of A onto M such that if ψ is a homomorphism of A into $N \in \mathbf{K}$, then there is a homomorphism $\alpha : M \to N$ such that $\varphi\alpha = \psi$. Prove that a maximal homomorphic image is unique up to isomorphism if it exists.

42. Starting with an arbitrary partial algebra $\langle A; \wedge, \vee \rangle$, carry out the construction of Theorem 5 (Theorem 24).

43. Find examples of posets $P_0 \subseteq P$ such that $F(P_0^m) = F(P^m)$.

44. After M. M. Gluhov [1960], a finite partial lattice $\mathfrak{A}$ is a *basis* of a lattice L if $L = F(\mathfrak{A})$, but $L = F(\mathfrak{A}_0)$ for no $A_0 \subseteq A$. Show that the lattice $L = F(\mathfrak{A})$ has more than one basis where $\mathfrak{A}$ is defined as follows: Let $\langle A; \leq \rangle$ be the poset given by Figure 5.11; for $x \leq y$ let $x \wedge y = y \wedge x = x$; furthermore, the join of any two elements is defined in $\{0, a, b, c, d, e, f, 1\}$ as supremum, and $1 \vee g = g \vee 1 = 1$. (This example, which is due to C. Herrmann, contradicts M. M. Gluhov's result.)

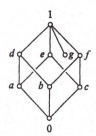

Figure 5.11

6. Special Elements

A *zero* of a poset P is an element 0 with $0 \leq x$ for all $x \in P$. A *unit*, 1, satisfies $x \leq 1$ for all $x \in P$. There are at most one zero and at most one unit. A *bounded* poset is one that has both 0 and 1.

A $\{0, 1\}$-*homomorphism* (of a bounded lattice into another one) is a homomorphism taking zero into zero and unit into unit. A $\{0, 1\}$-*sublattice* of a bounded lattice L is a sublattice containing the 0 and 1 of L. Similarly, we can define {0}-*homomorphism*, and so on, for lattices and semilattices.

In a bounded lattice L, a is a *complement* of b if $a \wedge b = 0$ and $a \vee b = 1$.

LEMMA 1.   *In a distributive lattice, an element can have only one complement.*

PROOF.   If b_0 and b_1 are both complements of a, then $b_0 = b_0 \wedge 1 = b_0 \wedge (a \vee b_1) = (b_0 \wedge a) \vee (b_0 \wedge b_1) = 0 \vee (b_0 \wedge b_1) = b_0 \wedge b_1$; similarly, $b_1 = b_0 \wedge b_1$, thus $b_0 = b_1$. ●

Let $a \in [b, c]$; x is a *relative complement* of a in $[b, c]$ if $a \wedge x = b, a \vee x = c$.

LEMMA 2.   *In a distributive lattice, if a has a complement, then it also has a relative complement in any interval containing it.*

PROOF.   Let d be the complement of a. Then $x = (d \vee b) \wedge c$ is the relative complement of a in $[b, c]$, provided that $b \leq a \leq c$. Indeed,

$$a \wedge x = a \wedge (d \vee b) \wedge c = ((a \wedge d) \vee (a \wedge b)) \wedge c$$
$$= (0 \vee b) \wedge c = b,$$
$$a \vee x = a \vee ((d \vee b) \wedge c) = (a \vee d \vee b) \wedge (a \vee c)$$
$$= 1 \wedge (a \vee c) = c. ●$$

LEMMA 3 (DE MORGAN'S IDENTITIES).   *In a distributive lattice, if a, b have complements a' and b', respectively, then $a \wedge b$ and $a \vee b$ have complements $(a \wedge b)'$ and $(a \vee b)'$, respectively, and*

$$(a \wedge b)' = a' \vee b';$$
$$(a \vee b)' = a' \wedge b'.$$

PROOF.   By Lemma 1 it suffices to prove that

$$(a \wedge b) \wedge (a' \vee b') = 0, \qquad (a \wedge b) \vee (a' \vee b') = 1$$

to verify the first law; the second is dual. Compute:

$$(a \wedge b) \wedge (a' \vee b') = (a \wedge b \wedge a') \vee (a \wedge b \wedge b')$$
$$= 0 \vee 0 = 0$$

and

$$(a \wedge b) \vee (a' \vee b') = (a \vee a' \vee b') \wedge (b \vee a' \vee b')$$
$$= 1 \wedge 1 = 1. \quad \bullet$$

A *complemented lattice* is a bounded lattice in which every element has a complement. A *relatively complemented lattice* is a lattice in which every element has a relative complement in any interval containing it. A *Boolean lattice B* is a complemented distributive lattice. Thus, in a Boolean lattice B, every element a has a unique complement, and B is also relatively complemented.

A *Boolean algebra* is a Boolean lattice in which 0, 1, and ' (complementation) are also considered to be operations. Thus a Boolean algebra is a system: $\langle B; \wedge, \vee, ', 0, 1 \rangle$, where $\wedge, \vee$ are binary, ' is a unary operation, and 0, 1 are nullary operations. (A nullary operation picks out an element of B.) A *homomorphism* φ preserves 0, 1 and '; that is, it is a $\{0, 1\}$-homomorphism satisfying $(x\varphi)' = x'\varphi$. A *subalgebra* is a $\{0, 1\}$-sublattice closed under '. $\mathfrak{B}_2$ will denote the two-element Boolean algebra.

Note that in a bounded distributive lattice L, if b is a complement of a, then b is the largest element x of L with $a \wedge x = 0$. More generally, let L be a lattice with 0. An element a^* is a *pseudocomplement* of a $(\in L)$ if $a \wedge a^* = 0$, and $a \wedge x = 0$ implies that $x \leq a^*$. An element can have at most one pseudocomplement. A *pseudocomplemented lattice* is one in which every element has a pseudocomplement.

The concept of pseudocomplement involves only the meet operation. Thus we can also define *pseudocomplemented semilattices*.

THEOREM 4.   *Let L be a pseudocomplemented meet-semilattice, $S(L) = \{a^* \mid a \in L\}$. Then the partial ordering of L partially orders $S(L)$ and makes $S(L)$ into a Boolean lattice. For $a,b \in S(L)$ we have $a \wedge b \in S(L)$, and the join in $S(L)$ is described by*

$$a \vee b = (a^* \wedge b^*)^*.$$

REMARK.   Even if L is a lattice, the join in L need not be the same as the join in $S(L)$. This result was proved for complete distributive lattices by V. Glivenko [1929]. In 1959 the result was established in its full generality by E. T. Schmidt and the author as a method of constructing lattices from semilattices. Since the intended application (to congruence lattices of lattices) fell through, the result was not published. The same result was rediscovered and published by O. Frink [1962]. Both proofs used special axiomatizations of Boolean algebras to get around the difficulty of proving distributivity. The present proof is direct and is the simplest of the three. E. T. Schmidt [1968] has succeeded in applying Theorem 4 as it was originally intended.

PROOF.   We start with the following observations:

$$a \leq a^{**}. \tag{1}$$

$$a \leq b \text{ implies that } a^* \geq b^*. \tag{2}$$

$$a^* = a^{***}. \tag{3}$$

$$a \in S(L) \text{ iff } a = a^{**}. \tag{4}$$

$$a,b \in S(L) \text{ implies that } a \wedge b \in S(L). \tag{5}$$

$$\text{For } a,b \in S(L), \sup_{S(L)} \{a, b\} = (a^* \wedge b^*)^*. \tag{6}$$

Formulas (1) and (2) follow from the definitions.

Formulas (1) and (2) yield $a^* \geq a^{***}$, and by (1) $a^* \leq a^{***}$, thus (3).

If $a \in S(L)$, then $a = b^*$; therefore, by (3), $a^{**} = b^{***} = b^* = a$. Conversely, if $a = a^{**}$, then $a = b^*$ with $b = a^*$; thus $a \in S(L)$, proving (4).

If $a,b \in S(L)$, then $a = a^{**}$, $b = b^{**}$, and so $a \geq (a \wedge b)^{**}$ and $b \geq (a \wedge b)^{**}$, thus $a \wedge b \geq (a \wedge b)^{**}$; by (1), $a \wedge b = (a \wedge b)^{**}$, thus $a \wedge b \in S(L)$. If $x \in S(L)$, $x \leq a$, and $x \leq b$, then $x \leq a \wedge b$; therefore, $a \wedge b = \inf_{S(L)} \{a, b\}$, proving (5).

$a^* \geq a^* \wedge b^*$, thus by (2) and (4), $a \leq (a^* \wedge b^*)^*$; similarly, $b \leq (a^* \wedge b^*)^*$. If $a \leq x$, $b \leq x$ ($x \in S(L)$), then $a^* \geq x^*$, $b^* \geq x^*$ by (2); thus by (2) and (4), $(a^* \wedge b^*)^* \leq x$, proving (6).

For $a,b \in S(L)$, define $a \vee b = (a^* \wedge b^*)^*$. By Formulas (5) and (6), $\langle S(L); \wedge, \vee \rangle$ is a bounded lattice. Since for $a \in S(L)$, $a \vee a^* = (a^* \wedge a^{**})^* = 0^* = 1$, $a \wedge a^* = 0$, $S(L)$ is a complemented lattice. Now we need only prove that $S(L)$ is distributive. For $x,y,z \in S(L)$, $x \wedge z \leq x \vee (y \wedge z)$ and $y \wedge z \leq x \vee (y \wedge z)$; therefore, $x \wedge z \wedge (x \vee (y \wedge z))^* = 0$ and $y \wedge z \wedge (x \vee (y \wedge z))^* = 0$. Thus $z \wedge (x \vee (y \wedge z))^* \leq x^*$ and y^*, and so $z \wedge (x \vee (y \wedge z))^* \leq x^* \wedge y^*$. Consequently, $z \wedge (x \vee (y \wedge z))^* \wedge (x^* \wedge y^*)^* = 0$, which implies that $z \wedge (x^* \wedge y^*)^* \leq (x \vee (y \wedge z))^{**}$. Now the left-hand side is $z \wedge (x \vee y)$ by Formula (6), and the right-hand side is $x \vee (y \wedge z)$ by Formula (4). Thus we get $z \wedge (x \vee y) \leq x \vee (y \wedge z)$, which is distributivity by Lemma 4.10. ●

Other types of special elements: An element a is an *atom* if $a \succ 0$ and a *dual atom* if $a \prec 1$; it is *join-irreducible* if $a = b \vee c$ implies that $a = b$ or $a = c$; it is *meet-irreducible* if $a = b \wedge c$ implies that $a = b$ or $a = c$. Examples are given in the following exercises.

Exercises

1. Find a homomorphism of bounded lattices that is not a $\{0, 1\}$-homomorphism and a sublattice that is not a $\{0, 1\}$-sublattice.
2. Find a modular lattice in which every element $x \neq 0, 1$ has exactly $\mathfrak{m}$ complements.
3. Let L be a distributive lattice, $a, b \in L$. Prove that if $a \vee b$ and $a \wedge b$ have complements, so do a and b.
4. Show that Lemma 2 holds in any modular lattice.
5. In a bounded lattice L, let x be a relative complement of a in $[b, c]$; let y be a relative complement of c in $[x, 1]$; let z be a relative complement of b in $[0, x]$; and let t be a relative complement of x in $[z, y]$. Verify that t is a complement of a.
6. Let B_0, B_1 be Boolean algebras and let φ be a $\{0, 1\}$-(lattice) homomorphism of B_0 into B_1. Show that φ is a homomorphism of the Boolean algebras.
7. Let L be a distributive lattice with 0. Show that $I(L)$ is pseudocomplemented.
8. Is the converse of exercise 7 true?
9. Prove that if, in Theorem 4, L is a lattice, then $a \vee b$ in $S(L)$ can be described by $a \vee b = (a \vee b)^{**}$.
10. Let L be a pseudocomplemented lattice. Show that $a^{**} \vee b^{**} = (a \vee b)^{**}$.
11. Find arbitrarily large pseudocomplemented lattices in which $S(L) = \{0, 1\}$.
12. Prove that in a Boolean lattice, $x \neq 0$ is join-irreducible iff x is an atom.
13. Show that in a finite lattice every element is the join of join-irreducible elements.
14. Verify that "finite lattice" in exercise 13 can be replaced by "lattice satisfying the *Descending Chain Condition*." (A lattice L or, in general, a poset, satisfies the Descending Chain Condition iff $x_0, x_1, x_2, \ldots \in L$, $x_0 \geq x_1 \geq x_2 \geq \cdots$ implies that $x_n = x_{n+1} = \cdots$ for some n; see exercise 2.7).
15. Show that some form of the Axiom of Choice must be used to verify exercise 14.
16. Prove that if a lattice satisfies the Descending Chain Condition, then every nonzero element contains an atom.
17. The dual of the Descending Chain Condition is the Ascending Chain Condition (see exercise 2.6). Dualize exercises 13–16.

18. Show that a lattice satisfies the Ascending Chain Condition and the Descending Chain Condition iff all chains are finite.
19. Find a lattice in which all chains are finite but that contains a chain of n elements for every natural number n.
20. Find a lattice in which there are no join- or meet-irreducible elements.
21. Prove that the Ascending Chain Condition (Descending Chain Condition) holds in a lattice L iff every ideal (dual ideal) of L is principal.
22. Show that the Ascending Chain Condition (Descending Chain Condition) holds in a poset P, then every element is contained in a maximal element (contains a minimal element).
23. Let $\wedge$ be a binary operation on L, let $*$ be a unary operation on L (that is, for every $\cdot a \in L$, $a^* \in L$), and let 0 be a nullary operation (that is, $0 \in L$). Let us assume that the following hold for all $a,b,c \in L$: $a \wedge b = b \wedge a$, $(a \wedge b) \wedge c = a \wedge (b \wedge c)$, $a \wedge a = a$, $0 \wedge a = 0$, $a \wedge (a \wedge b)^* = a \wedge b^*$, $a \wedge 0^* = a$, $(0^*)^* = 0$. Show that $\langle L; \wedge \rangle$ is a meet-semilattice with 0 as zero, and for all $a \in L$, a^* is the pseudo-complement of a (R. Balbes and A. Horn [1970a]).
24. Let L be a pseudocomplemented meet-semilattice and let $a,b \in L$. Verify the formulas $(a \wedge b)^* = (a^{**} \wedge b)^* = (a^{**} \wedge b^{**})^*$.
25. Let L be a meet-semilattice and let $a,b \in L$. The *pseudocomplement* $a * b$ *of a relative to b* is an element of L satisfying $a \wedge x \leq b$ iff $x \leq a * b$. Show that $a * b$ is unique if it exists; show that $a * a$ exists iff L has a unit.
26. Let L be a *relatively pseudocomplemented meet-semilattice* (that is, L is a meet-semilattice and $a * b$ exists for all $a,b \in L$). Show that L has 0 and 1; for all $a,b,c \in L$: $a * (b * c) = (a \wedge b) * c$ and $a * (b * c) = (a * b) * (a * c)$. Furthermore, if L is a lattice, then L is distributive.
27. Let L be a pseudocomplemented distributive lattice. Prove that for each $a \in L$, $(a]$ is a pseudocomplemented distributive lattice; in fact, the pseudocomplement of $x \in (a]$ in $(a]$ is $x^* \wedge a$.
*28. Using the notation of exercise 27, let $S(a)$ denote the elements of the form $x^* \wedge a$, $x \leq a$. Then $S(a)$ is a Boolean algebra by Theorem 4. Let $\vee_a$ denote the join in $S(a)$. Show that if $x,y \in S(a)$ and $x,y \in S(b)$ $(a,b \in L)$, then $x \vee_a y = x \vee_b y$.
*29. Let $b \in S(a)$. Prove that $S(b) \subseteq S(a)$. (The results of exercises 28 and 29 appear to be new.)
30. Show that T_n (see exercise 2.36) is complemented.
31. Show that in T_n every interval is pseudocomplemented. (Exercises 30 and 31 are due to H. Lakser.)

Further Topics and References

Many of the concepts and results discussed in Chapter 1 are special cases of universal algebraic concepts and results. To see this, the reader needs the definition of a universal algebra. An *n-ary operation* f on a set A is a map

from A^n into A; in other words, if $a_1, \ldots, a_n \in A$, then $f(a_1, \ldots, a_n) \in A$.
If $n = 1$, f is called *unary*; if $n = 2$, f is called *binary*. Since $A^0 = \{\varnothing\}$,
a *nullary operation* ($n = 0$) is determined by $f(\varnothing) \in A$, and f is sometimes
identified with $f(\varnothing)$. A *universal algebra*, or simply *algebra*, consists of a
nonvoid set A and a set F of operations; each $f \in F$ is an n-ary operation
for some n (depending on f). We denote this algebra by $\mathfrak{A}$ or $\langle A; F \rangle$.
Many of the results of Sections 3–5 can be formulated and proved for
arbitrary universal algebras. (We shall utilize this fact in Chapter 3.) For
more details, see Chapters 1–4 of the author's book [1968].

In every poset we can introduce (as suggested by the real line) a ternary
relation r called *betweenness*: $r(a, b, c)$ iff $a \leq b \leq c$ or $c \leq b \leq a$.
M. Altwegg [1950] proved that the partial ordering can be defined in
terms of betweenness (naturally, up to duality).

Lattices and Boolean algebras can be defined by identities in innumer-
able ways. Of the eight identities we used to define lattices ((L1)–(L4) of
Section 1), two ((L1)) can be dropped. Ju. I. Sorkin [1951] reduced six to
four, R. Padmanabhan [1969] later found two, and finally R. N. McKenzie
[a] found a single identity characterizing lattices. Ju. I. Sorkin's identities
use only three variables; the others use more. It is easy to see that two
variables would not suffice. Take the lattice of Figure 5.8 and redefine the
join of the two atoms to be 1; otherwise keep all the joins and meets.
The resulting algebra is not a lattice, but every subalgebra generated by
two elements is a lattice. Therefore, lattices cannot be defined by identities
in two variables. By means of a more complicated construction, A. H.
Diamond and J. C. C. McKinsey [1947] derived the same conclusion for
Boolean algebras.

A recent result of A. Tarski [1968] states that, given any integer n, there
exist n identities defining lattices such that no identity can be dropped.

By exercise 4.24, modular and distributive lattices can be defined by
two identities. Nicer sets of identities for these cases were found by
M. Kolibiar [1956] $((a \vee (b \wedge b)) \wedge b = b, ((a \wedge b) \wedge c) \vee (a \wedge d) =
((d \wedge a) \vee (c \wedge b)) \wedge a$ characterize modular lattices) and M. Sholander
[1951] $(a \wedge (a \vee b) = a, a \wedge (b \vee c) = (c \wedge a) \vee (b \wedge a)$ characterize
distributive lattices). See also J. Riečan [1958].

One of the most useful axiomatizations of Boolean algebras is that of
E. V. Huntington [1904],[2] according to which a Boolean algebra is a complemented lattice in which the complementation is pseudocomplemen-

---

[2] Observe that a proof of Huntington's result is implicit in the proof of Theorem
6.4.

tation (that is, $a \wedge x = 0$ implies that $x \leq a'$). It had been long conjectured that every complemented lattice with unique complements was Boolean; this conjecture was disproved by R. P. Dilworth [1945]. One of the briefest axiom systems of Boolean algebras in terms of $\wedge$ and $'$ is that of L. Byrne [1946]: $a \wedge b = b \wedge a, a \wedge (b \wedge c) = (a \wedge b) \wedge c, a \wedge b' = c \wedge c'$ iff $a \wedge b = a$. Characterization by identities is usually longer. The observation has been made only recently (independently, by R. N. McKenzie, A. Tarski, and the author) that Boolean algebras can be defined by a single identity (see A. Tarski [1968], and G. Grätzer and R. N. McKenzie [1967]). A thorough survey of the axiom systems of Boolean algebras is given in S. Rudeanu [1963]; see also F. M. Sioson [1964].

Identities for semilattices are easy to provide. Two identities characterize semilattices (D. H. Potts [1965] and R. Padmanabhan [1966]); one identity is not enough (D. H. Potts [1965]).

A diagram of a finite lattice is a *graph*; however, very little work has been done in lattice theory from a graph theoretic point of view. It is known that a finite distributive lattice is planar iff no element is covered by more than two elements. Pairs of lattices whose diagrams are isomorphic as graphs were investigated by J. Jakubik [1954a], [1954b]. See also J. Jakubik and M. Kolibiar [1954].

The examples of lattice identities we have seen so far seem to suggest that all such identities are self-dual. The identity $(a \wedge b) \vee (a \wedge c) = a$, where $a = x \wedge ((x \wedge y) \vee (y \wedge z) \vee (z \wedge x))$, $b = (x \wedge y) \vee (y \wedge z)$, $c = (x \wedge z) \vee (y \wedge z)$, is an example of a nonself-dual identity (H. F. J. Lowig [1943]).

Sublattices suggest a number of problems: For an equational class **K** of lattices, let $f_{\mathbf{K}}(k, n)$ denote the smallest integer such that any lattice in **K** having at least $f_{\mathbf{K}}(k, n)$ elements has at least k sublattices of exactly n elements; write $f(k, n)$ for $f_{\mathbf{L}}(k, n)$. The function $f(1, n)$ has been investigated: $f(1, t) = t$ for $t \leq 6$ (I. Kaplansky); $f(1, n) \leq n^{4n-3}$ (R. Wille, unpublished); and $f(1, n) \leq n^{3n}$ (G. Havas and M. Ward [1969]); see also M. Curzio [1953]. The lattice of sublattices Sub (L) and subsemilattices of a semilattice were investigated by N. D. Filippov [1966] and L. N. Ševrin [1964], respectively. They gave necessary and sufficient conditions for Sub $(L) \cong$ Sub (L_1). Filippov's results showed that if L is distributive (Boolean) and Sub $(L) \cong$ Sub (L_1), then L_1 is distributive (Boolean). The lattice of subalgebras of a Boolean algebra has been characterized in D. Sachs [1962].

If K is a sublattice of L, then K is isomorphic to a sublattice of $I(L)$; finite lattices K for which the reverse holds were investigated by G. Grätzer [1970].

An *endomorphism* φ ($\{0, 1\}$-*endomorphism*) of the lattice L is a homomorphism ($\{0, 1\}$-homomorphism) of L into itself. If, in addition, φ is one-to-one and onto, then φ is an *automorphism*. Let $E(L)$, $E_{0,1}(L)$, and $A(L)$ denote the set of all endomorphisms, $\{0, 1\}$-endomorphisms, and automorphisms of L, respectively. (We form $E_{0,1}(L)$ iff L has 0 and 1.) Defining the multiplication as composition of maps, $E(L)$ and $E_{0,1}(L)$ are *monoids* (semigroups with identity), and $A(L)$ is a group. G. Birkhoff [1946] has shown that for any group G, there exists a distributive lattice L with $A(L) \cong G$; if G is finite, so is L. Small lattices L with given $A(L)$ were investigated by R. Frucht [1948], [1950]. For any given monoid M, a bounded lattice L with $E_{0,1}(L) \cong M$ was constructed in G. Grätzer and J. Sichler [a]—see also R. N. McKenzie and J. Sichler [a] and J. Sichler [a].

A very important property of complete lattices is the fixed-point theorem: Any isotone map f of a complete lattice L into itself has a fixed point (that is, $f(a) = a$ for some $a \in L$); in fact, $f(a) = a$ if a is defined by

$$a = \bigvee (b \mid b \in L, b \leq f(b))$$

—see A. Tarski [1955]. A. C. Davis [1955] proved that if this theorem holds in a lattice L, then L is complete. Various generalizations are given in A. Tarski [1955], E. S. Wolk [1957], A. Pelczar [1961] and [1962], S. Abian and A. B. Brown [1961], V. Devidé [1963], S. R. Kogalovskiĭ [1964], and R. Demarr [1964].

Boolean algebras originated as an algebraic formalization of propositional logic. Most introductory logic texts give satisfactory expositions of these ideas; see also P. R. Halmos [1963]. There is a similar relationship between Boolean algebras and the theory of switching circuits; see M. A. Harrison [1965].

The construction of lattices described in exercise 4.20 was investigated by P. M. Whitman [1943].

It is hard to overemphasize the importance of free lattices; they provide one of the most important research tools of lattice theory. Two typical applications to modular lattices can be found in G. Grätzer and E. T. Schmidt [1961] and G. Grätzer [1966].

P. Ribenboim [1949] first pointed out that pseudocomplementation can be described by identities (involving *). A. Monteiro [1955] accomplished the same result for relative pseudocomplementation; see also R. Balbes and A. Horn [1970a]. This fact is applied in G. Grätzer [1969].

The connections of lattice theory with topology are manifold and have been excluded from this book. The closed (and also the open) subsets of a topological space X form a lattice $L(X)$, and this lattice determines the topological space (provided it is T_1). These lattices were characterized by H. Wallman [1938]. Thus certain parts of topology can be studied within lattice theory—the theory of compactifications being one example (H. Wallman [1938], O. Frink [1964]). The set of all continuous real functions on a topological space X also forms a lattice $C(X)$ under pointwise ordering. The lattice $C(X)$ determines X if X is compact Hausdorff (I. Kaplansky [1947]; see also L. Gillman and M. Jerison [1960]).

All topologies on a set X form a lattice $T(X)$ if, for topologies S_1, S_2 on X, we set $S_1 \leq S_2$ whenever every S_1-open set is also S_2-open. This point of view permeates the topology book of R. Vaidyanathaswamy [1947] and has influenced some parts of modern topology. A typical recent result is that of A. K. Steiner [1966], according to which $T(X)$ is a complemented lattice.

A *topological lattice L* is a lattice equipped with a topology such that $\wedge$ and $\vee$ are continuous functions. The theory of topological lattices, the study of which was started by A. D. Wallace and his students, is quite extensive. The methods used are mostly topological rather than lattice theoretical in nature. The reader should consult L. W. Anderson [1961] for an early review of this field.

In contrast with topological lattices, intrinsic topologies are considered on arbitrary lattices. *Intrinsic topologies* are topologies defined on a lattice in terms of the partial ordering. For instance, the interval topology takes the intervals as a subbase for closed sets. For more details, see O. Frink [1942], G. Birkhoff [1962], and B. C. Rennie [1951].

Few of the topics in this chapter have been investigated for semilattices. Congruences are one exception (see G. Zacher [1952], D. Papert [1964], R. A. Dean and R. H. Oehmke [1964], R. Permutti [1964], J. Varlet [1965], and E. T. Schmidt [1969]). Congruences of partial lattices are considered in G. Grätzer and H. Lakser [1968].

PROBLEMS

1. Characterize the endomorphism semigroup of a lattice.
2. Characterize $E(L)$ and $E_{0,1}(L)$ for finite lattices and for lattices of finite length (see J. Sichler [a]).[3]
3. For an arbitrary monoid M, find a lattice L such that the nonconstant endomorphisms of L, $E_n(L)$ form a semigroup, and $E_n(L) \cong M$.[4]
4. Characterize the endomorphism semigroups of complemented lattices. (The categorical versions of problems 1–4 should also be considered.)
5. Characterize the comparability relation of semilattices and of lattices (see exercises 1.34 and 1.35).[5]
6. Characterize the lattices T_n and finite lattices that can be embedded in some T_n (see exercises 2.26–2.36, 6.30, and 6.31).
7. Characterize the lattice Sub (L) of all sublattices of a lattice L.
8. For what equational classes **K** does $L \in \mathbf{K}$, Sub $(L) \cong$ Sub (L_1) imply that $L_1 \in K$?
9. Characterize planar lattices.
10. Define the *complexity* of a finite lattice L as the number of intersecting lines in an optimal diagram. Investigate this concept.
11. For an equational class **K**, which integers n can occur as the complexity of a finite lattice $L \in \mathbf{K}$?
12. Let $L = F_{\mathbf{K}}(A)$, where **K** is an equational class and A is a *finite* partial lattice; L is then called *finitely presented over* **K**. Is the automorphism group of L necessarily finite? Is this true at least if $\mathbf{K} = \mathbf{L}$?
13. Does a finitely presented lattice have only finitely many bases in the sense of exercise 5.44? Is this true at least if $\mathbf{K} = \mathbf{L}$?
14. For which finite lattices K does it hold that if K is a sublattice of $I(L)$, then K is a sublattice of L (see G. Grätzer [1970])?
15. Compare finite lattices K satisfying the condition of problem 14 with those (i) that can be embedded in $F(3)$, or (ii) for which $\{N \mid N \in \mathbf{L}, K$ is not a sublattice of $N\}$ is equational.
16. Describe the equational class of lattices with $n \leq 3$ identities involving only three variables.
17. Find the shortest single identity characterizing lattices.
18. Does there exist a non-Boolean finite algebra $\langle A; \wedge, \vee, {}', 0, 1 \rangle$ such that every two-generated subalgebra is Boolean (see A. H. Diamond and J. C. C. McKinsey [1947])?

---

[3] R. N. McKenzie proved that any $E(L)$ can be represented as $E_{0,1}(L_1)$ and that if L is finite or of finite length, then so is L_1.

[4] This problem was solved by R. N. McKenzie using the General Continuum Hypothesis. Subsequently, J. Sichler solved it in the categorical form without the Hypothesis.

[5] R. N. McKenzie has shown that neither of the two problems has a first-order solution.

19. Define and investigate weak partial lattices and partial lattices satisfying an identity.

20. For a partial lattice A, define a *congruence relation* as follows: A binary relation Θ on A is a congruence if there exists a lattice L and a congruence relation Φ on L such that (i) A is a relative sublattice of L and (ii) Φ restricted to A is Θ. Give an intrinsic characterization of this concept of congruence relation.[6]

21. Characterize the lattice of all congruence relations of a semilattice.

22. Let L be a meet-semilattice with 0 for which $[0, a]$ is pseudocomplemented for each $a \in L$; let $S(a)$ denote the pseudocomplements in $[0, a]$. Characterize the family of Boolean algebras $\{S(a) \mid a \in L\}$ (see exercises 6.28 and 6.29).

23. What is the order of magnitude of $f(k, n)$? (Note that $f(k, n)$ is defined in Further Topics and References.)[7]

24. For what equational classes K do all $f_K(k, n)$ exist?

25. Give a concrete example of a non-Boolean uniquely complemented lattice (see R. P. Dilworth [1945]; C. C. Chen and G. Grätzer [1969a]).

26. Does there exist a complete non-Boolean uniquely complemented lattice?

27. Investigate the lattice of join-endomorphisms of a finite lattice (see G. Grätzer and E. T. Schmidt [1958a]).

---

[6] This is known for universal algebras in general; see G. Grätzer [1968], Theorem 13.3.

[7] B. Wolk pointed out that the argument of G. Havas and M. Ward [1969] can be modified so as to prove that if L has more than n^{3^n} elements, then L has an n element sublattice in which all but at most two elements are join- and meet-irreducible. This implies the existence of $f(k, n)$.

CHAPTER 2

DISTRIBUTIVE LATTICES

7. Characterization Theorems and Representation Theorems

The two typical examples of nondistributive lattices are $\mathfrak{M}_5$ and $\mathfrak{N}_5$, whose diagrams are given in Figure 7.1.

Theorem 1 is a striking and useful characterization of distributive lattices; Theorem 2 is a more detailed version of Theorem 1.

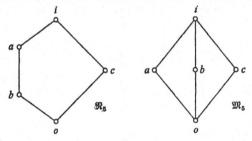

Figure 7.1

THEOREM 1.   *A lattice L is distributive iff L has no sublattice isomorphic to $\mathfrak{M}_5$ or $\mathfrak{N}_5$.*

THEOREM 2.
(i) *A lattice L is modular iff it has no sublattice isomorphic to $\mathfrak{N}_5$.*
(ii) *A modular lattice L is distributive iff it has no sublattice isomorphic to $\mathfrak{M}_5$.*

PROOF.
(i) If L is modular, then every sublattice of L is also modular; $\mathfrak{N}_5$ is not modular, thus it cannot be isomorphic to a sublattice of L. Conversely, let L be nonmodular, let $a,b,c \in L$, $a \geq b$, and let $(a \wedge c) \vee b \neq a \wedge (c \vee b)$. The free lattice generated by a, b, and c with $a \geq b$ is shown in Figure 2.3. Therefore, the sublattice of L generated by a, b, and c must be a homomorphic image of the lattice of Figure 2.3. Observe that if any two of the five elements $a \wedge c$, $(a \wedge c) \vee b$, $a \wedge (b \vee c)$, $b \vee c$, c are identified under a homomorphism, then so are $(a \wedge c) \vee b$ and $a \wedge (b \vee c)$. Consequently, these five elements are distinct in L, and they form a sublattice isomorphic to $\mathfrak{N}_5$. ▶

(ii) Let L be modular, but nondistributive, and choose $x,y,z \in L$ such that $x \wedge (y \vee z) \neq (x \wedge y) \vee (x \wedge z)$. The free modular lattice generated by x, y, z is shown in Figure 5.7. By inspecting the diagram we see that u, x_1, y_1, z_1, v form a sublattice isomorphic to $\mathfrak{M}_5$. Thus in *any* modular lattice they form a sublattice isomorphic to a quotient lattice of $\mathfrak{M}_5$. But $\mathfrak{M}_5$ has only two quotient lattices: $\mathfrak{M}_5$ and the one-element lattice. In the former case we have finished the proof. In the latter case, note that if u and v collapse, then so do $x \wedge (y \vee z)$ and $(x \wedge y) \vee (x \wedge z)$, contrary to our assumption. ●

Naturally, these results could be proved without any reference to free lattices. A routine proof of (ii) runs as follows: Take $x,y,z \in L$ such that $x \wedge (y \vee z) \neq (x \wedge y) \vee (x \wedge z)$ and define u, v, x_1, y_1, z_1 as the corresponding polynomials of Figure 5.7. Then a direct computation shows that u, v, x_1, y_1, z_1 form a sublattice isomorphic to $\mathfrak{M}_5$. There is a very natural objection to such a proof, namely: How are the appropriate polynomials found? How is it possible to guess the result? And there is only one answer: by working out the free lattice.

COROLLARY 3.   *A lattice L is distributive iff every element has at most one relative complement in any interval.*

PROOF. The "only if" part was proved in Section 6. If L is nondistributive, then by Theorem 2 it contains $\mathfrak{N}_5$ or $\mathfrak{M}_5$, and each has an element with two relative complements in some interval. ⬤

COROLLARY 4.   *A lattice L is distributive iff for any two ideals I, J of L:*

$$I \vee J = \{i \vee j \mid i \in I, j \in J\}.$$

PROOF. Let L be distributive. By Lemma 3.1(ii), if $t \in I \vee J$, then $t \leq i \vee j$ for some $i \in I$, $j \in J$. Therefore, $t = (t \wedge i) \vee (t \wedge j)$, $t \wedge i \in I$, $t \wedge j \in J$. Conversely, if L is nondistributive, then L contains elements a, b, c as in Figure 7.1. Let $I = (b]$ and $J = (c]$; observe that $a \in I \vee J$, since $a \leq b \vee c$. However, a has no representation as required by Corollary 4, since if $a = i \vee j$, $i \in I$, $j \in J$, then $j \leq a$, $j \leq c$. Therefore, $j \leq a \wedge c < b$, thus $j \in I$, and so $a = i \vee j \in I$, a contradiction. ⬤

Another important property of ideals of a distributive lattice is:

LEMMA 5.   *Let I and J be ideals of a distributive lattice L. If $I \wedge J$ and $I \vee J$ are principal, then so are I and J.*

PROOF. Let $I \wedge J = (x]$ and $I \vee J = (y]$. Then $y = i \vee j$ for some $i \in I$, $j \in J$. Set $c = x \vee i$ and $b = x \vee j$; note that $c \in I$, $b \in J$. We claim that $I = (c]$, $J = (b]$. Indeed, if, for instance, $J \neq (b]$, then there is an $a > b$, $a \in J$, and $\{x, a, b, c, y\}$ form an $\mathfrak{N}_5$. ⬤

THEOREM 6.   *Let L be a distributive lattice and let $a \in L$. Then the map*

$$\varphi: x \to \langle x \wedge a, x \vee a \rangle, \qquad x \in L$$

is an embedding of L into $(a] \times [a)$; it is an isomorphism if a has a complement.

PROOF. φ is one-to-one, since if $x\varphi = y\varphi$, then x and y are both relative complements of a in the same interval; thus $x = y$ by Corollary 3. Distributivity implies also that φ is a homomorphism.

If a has a complement, b, and $\langle u, v \rangle \in (a] \times [a)$, then for $x = (u \vee b) \wedge v$, $x\varphi = \langle u, v \rangle$; therefore, φ is an isomorphism. ⬤

We start the detailed investigation of the structure of distributive lattices with the finite case.

DEFINITION 7. *For a distributive lattice, L, let J(L) denote the set of all nonzero join-irreducible elements, regarded as a poset under the partial ordering of L. For $a \in L$ set*

$$r(a) = \{x \mid x \leq a, x \in J(L)\}.$$

DEFINITION 8. *For a poset P, call $A \subseteq P$ hereditary if $x \in A$, $y \leq x$ implies that $y \in A$. Let $H(P)$ denote the set of all hereditary subsets partially ordered by set inclusion.*

Note that $H(P)$ is a lattice in which meet and join are intersection and union, respectively, and thus $H(P)$ is distributive. The structure of finite distributive lattices is revealed by the following result:

THEOREM 9. *Let L be a finite distributive lattice. Then the map*

$$\varphi: a \to r(a)$$

is an isomorphism between L and $H(J(L))$.

PROOF. Since L is finite, every element is the join of join-irreducible elements; thus

$$a = \bigvee r(a),$$

showing that φ is one-to-one. Obviously, $r(a) \cap r(b) = r(a \wedge b)$, and so $(a \wedge b)\varphi = a\varphi \wedge b\varphi$. The formula $(a \vee b)\varphi = a\varphi \vee b\varphi$ is equivalent to

$$r(a \vee b) = r(a) \cup r(b).$$

To verify this formula, note that $r(a) \cup r(b) \subseteq r(a \vee b)$ is trivial. Now let $x \in r(a \vee b)$. Then $x = x \wedge (a \vee b) = (x \wedge a) \vee (x \wedge b)$; therefore, $x = x \wedge a$ or $x = x \wedge b$, since x is join-irreducible. Thus $x \in r(a)$ or $x \in r(b)$, that is, $x \in r(a) \cup r(b)$.

Finally, we have to show that if $A \in H(J(L))$, then $a\varphi = A$ for some $a \in L$. Set $a = \bigvee A$. Then $r(a) \supseteq A$ is obvious. Let $x \in r(a)$; then $x = x \wedge a = x \wedge \bigvee A = \bigvee (x \wedge y \mid y \in A)$. So $x = x \wedge y$ for some $y \in A$, implying that $x \in A$, since A is hereditary. ●

COROLLARY 10. *The correspondence $L \to J(L)$ makes the class of all finite distributive lattices correspond to the class of all finite posets; isomorphic lattices correspond to isomorphic posets, and vice versa.*

PROOF. This is obvious from $J(H(P)) \cong P$, and $H(J(L)) \cong L$. ●

A subset S of $P(A)$ is called a *ring of sets* if $X, Y \in S$ implies that $X \cap Y$, $X \cup Y \in S$. Since $H(J(L))$ is a ring of sets, we obtain

COROLLARY 11. *A finite lattice is distributive iff it is isomorphic to a ring of sets.*

If the poset Q is unordered, $H(Q) = P(Q)$; if B is Boolean, $J(B)$ is the set of all atoms, and therefore $J(B)$ is unordered. Thus we get:

COROLLARY 12. *A finite lattice is Boolean iff it is isomorphic to the Boolean lattice of all subsets of a finite set.*

A representation $a = x_0 \vee \cdots \vee x_{n-1}$ is *redundant* if

$$a = x_0 \vee \cdots \vee x_{i-1} \vee x_{i+1} \vee \cdots \vee x_{n-1},$$

for some $0 \le i < n$; otherwise it is *irredundant*.

COROLLARY 13. *Every element of a finite distributive lattice has a unique irredundant representation as a join of join-irreducible elements.*

PROOF. The existence of such a representation is obvious. If

$$a = x_0 \vee \cdots \vee x_{n-1}$$

is an irredundant representation, then $r(a) = \bigcup (r(x_i) \mid 0 \le i < n)$. Thus x occurs in such a representation iff x is a maximal element of $r(a)$; hence the uniqueness. ●

A finite chain of n elements is said to be of *length $n - 1$*.

COROLLARY 14. *Every maximal chain C of the finite distributive lattice L is of length $|J(L)|$.*

PROOF. For $a \in J(L)$, let $m(a)$ be the smallest member of C containing a. Then $a \to m(a)$ is a one-to-one map of $J(L)$ onto the nonzero elements of C. Indeed, if $m(a) = m(b)$, $m(a) \succ x$, and $x \in C$, then $x \vee a = x \vee b$; therefore, $a = (a \wedge x) \vee (a \wedge b)$, implying that $a \le x$ or $a \le b$. But $a \le x$ implies that $m(a) \le x < m(a)$, a contradiction. Consequently, $a \le b$; similarly, $b \le a$; thus $a = b$. Let $y \in C$, $y \succ z$, $z \in C$. Then $r(y) \supset r(z)$, and so $y = m(a)$ for any $a \in r(y) - r(z)$. ●

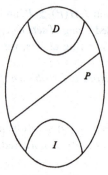

Figure 7.2

The crucial Theorem 9 and its most important consequence, Corollary 11, depend on the existence of sufficiently many join-irreducible elements. In an infinite distributive lattice there may be no join-irreducible element. Note that in a distributive lattice, a is join-irreducible iff $L - [a)$ is a prime ideal. In the infinite case the role of join-irreducible elements is taken over by prime ideals. The crucial result is the existence of sufficiently many prime ideals (as illustrated in Figure 7.2).

THEOREM 15   (M. H. Stone [1936]).   *Let L be a distributive lattice, let I be an ideal, let D be a dual ideal of L, and let $I \cap D = \varnothing$. Then there exists a prime ideal P of L such that $P \supseteq I$ and $P \cap D = \varnothing$.*

PROOF.   Some form of the Axiom of Choice is needed to prove this statement. The most convenient form for this proof is:

ZORN'S LEMMA.   *Let A be a set and let $\mathscr{X}$ be a nonvoid subset of $P(A)$. Let us assume that $\mathscr{X}$ has the following property: If $C \subseteq \mathscr{X}$ and C is a chain (that is, for any $X, Y \in C$ we have $X \subseteq Y$ or $Y \subseteq X$), then*

$$\bigcup (X \mid X \in C) \in \mathscr{X}.$$

Then $\mathscr{X}$ has a maximal member (that is, an $M \in \mathscr{X}$ such that $M \subseteq X$ and $X \in \mathscr{X}$ imply that $M = X$).

Let $\mathscr{X}$ be the set of all ideals of L that contain I and that are disjoint from D. We have to verify that $\mathscr{X}$ satisfies the hypothesis of Zorn's Lemma. Since $I \in \mathscr{X}$, we conclude that $\mathscr{X}$ is nonvoid. Let C be a chain in $\mathscr{X}$ and let

$M = \bigcup (X \mid X \in C)$. If $a,b \in M$, then $a \in X$, $b \in Y$ for some $X,Y \in C$; since C is a chain, either $X \subseteq Y$ or $Y \subseteq X$; if, say, $X \subseteq Y$, then $a,b \in Y$, and so $a \vee b \in Y \subseteq M$, since Y is an ideal. Also, if $b \leq a \in M$, then $a \in X$; since X is an ideal, $b \in X \subseteq M$. Thus M is an ideal. It is obvious that $M \supseteq I$ and $M \cap D = \varnothing$, verifying that $M \in \mathcal{X}$. Therefore, by Zorn's Lemma, $\mathcal{X}$ has a maximal element, say, P. We claim that P is a prime ideal. Indeed, if P is not prime, there exist $a,b \in L$ such that $a,b \notin P$ but $a \wedge b \in P$. Because of the maximality of $P, (P \vee (a]) \cap D \neq \varnothing, (P \vee (b]) \cap D \neq \varnothing$. Let $p \vee a \in D, q \vee b \in D, p,q \in P$. Then $x = (p \vee a) \wedge (q \vee b) \in D$, since D is a dual ideal. Also, $x = (p \wedge q) \vee (p \wedge b) \vee (a \wedge q) \vee (a \wedge b) \in P$; thus $P \cap D \neq \varnothing$, a contradiction. ●

COROLLARY 16. *Let L be a distributive lattice, let I be an ideal of L, and let $a \in L$ and $a \notin I$. Then there is a prime ideal P such that $P \supseteq I$, and $a \notin P$.*

PROOF. Apply Theorem 15 to I and $D = [a)$. ●

COROLLARY 17. *Let L be a distributive lattice, $a,b \in L$, $a \neq b$. Then there is a prime ideal containing exactly one of a and b.*

PROOF. Either $(a] \cap [b) = \varnothing$, or $[a) \cap (b] = \varnothing$. ●

COROLLARY 18. *Every ideal I of a distributive lattice is the intersection of all prime ideals containing it.*

PROOF. Let $I_1 = \bigcap (P \mid P \supseteq I, P$ is a prime ideal of $L)$. If $I \neq I_1$, then there is an $a \in I_1 - I$, and so by Corollary 16 there is a prime ideal P, with $P \supseteq I, a \notin P$. But then $a \notin P \supseteq I_1$ is a contradiction. ●

THEOREM 19 (G. Birkhoff [1933] and M. H. Stone [1936]). *A lattice is distributive iff it is isomorphic to a ring of sets.*

PROOF. Let L be a distributive lattice and let X be the set of all prime ideals of L. For $a \in L$ set

$$r(a) = \{P \mid a \notin P, P \in X\}.$$

Then the $r(a)$ form a ring of sets, and $a \to r(a)$ is an isomorphism. The details are similar to the proof of Theorem 9, except for the first step, which now uses Corollary 17. ●

This result has a very useful corollary.

COROLLARY 20. *Let L be a distributive lattice with more than one element. An identity holds in L iff it holds in the two-element chain, $\mathfrak{C}_2$.*

PROOF. Let $p = q$ hold in L. Since $|L| > 1$, $\mathfrak{C}_2$ is a sublattice of L, and $p = q$ holds in $\mathfrak{C}_2$. Conversely, let $p = q$ hold in $\mathfrak{C}_2$. Note that $\mathfrak{C}_2 = P(X)$, with $|X| = 1$, and that $P(A)$ is isomorphic to the direct power $P(X)^{|A|}$. Therefore, $p = q$ holds in any $P(A)$. By Theorem 19, L is a sublattice of some $P(A)$; thus $p = q$ holds in L. ●

We can reformulate Theorem 19 using the concept of a field of sets: A *field of sets* is a ring of sets closed under set complementation.

COROLLARY 21 (M. H. Stone [1936]). *A lattice is Boolean iff it is isomorphic to a field of sets.*

PROOF. Use the representation of Theorem 19. Obviously $r(a') = X - r(a)$, and thus complements are also preserved. ●

Let $\mathscr{P}(L)$ denote the set of all prime ideals of L, regarded as a poset under $\subseteq$. The importance of $\mathscr{P}(L)$ should be clear from the previous results. A topology on $\mathscr{P}(L)$ will be discussed in Section 11.

Some interesting properties of L are reflected in $\mathscr{P}(L)$. An important result of this type is the following theorem (see also L. Rieger [1949]):

THEOREM 22 (L. Nachbin [1947]). *Let L be a distributive lattice with 0 and 1. Then L is a Boolean lattice iff $\mathscr{P}(L)$ is unordered.*

PROOF. Let L be Boolean, $P, Q \in \mathscr{P}(L)$, $P \subset Q$. Choose $a \in Q - P$. Since $a \in Q$, $a' \notin Q$, and thus $a' \notin P$. Thus $a, a' \notin P$, but $a \wedge a' = 0 \in P$, a contradiction, showing that $\mathscr{P}(L)$ is unordered.[1]

Now let $\mathscr{P}(L)$ be unordered and $a \in L$, and let us assume that a has no complement. Set $D = \{x \mid a \vee x = 1\}$. Then D is a dual ideal. Take $D_1 = D \vee [a) = \{x \mid x \geq d \wedge a \text{ for some } d \in D\}$. The dual ideal D_1 does not contain 0, since $0 = d \wedge a$, $a \vee d = 1$ would mean that d is a complement of a. Thus there exists a prime ideal P disjoint to D_1. Note that $1 \notin (a] \vee P$, otherwise $1 = a \vee p$ for some $p \in P$, contradicting $P \cap D = \varnothing$.

---

[1] This proof in fact verifies that in a Boolean algebra every prime ideal is maximal.

Thus some prime ideal Q contains $(a] \vee P$; and so $P \subset Q$, which is impossible since $P(L)$ is unordered. ●

According to Corollary 18, every ideal is an intersection of primes. When is this representation unique?

THEOREM 23 (J. Hashimoto [1952]). *Let L be a distributive lattice with 0 and 1. Every ideal has a unique representation as an intersection of prime ideals iff L is a finite Boolean lattice.*

PROOF.   If L is a finite Boolean lattice, then P is a prime ideal iff $P = (a]$, where a is a dual atom; the uniqueness follows from Corollary 13 (or is obvious by direct computation).

Now let every ideal of L have a unique representation as a meet of prime ideals. We claim that $I(L)$ is Boolean. Let $I \in I(L)$,

$$J = \bigcap (P \mid P \in \mathscr{P}(L), P \nsupseteq I).$$

Then $I \wedge J = \bigcap (P \mid P \in \mathscr{P}(L)) = (0]$. If $L \neq I \vee J$, then there is a prime ideal $P_0 \supseteq I \vee J$, and consequently J has two representations:

$$\bigcap (P \mid P \nsupseteq I) = P_0 \cap \bigcap (P \mid P \nsupseteq I).$$

Thus $L = I \vee J$, and J is a complement of I in $I(L)$.

By Lemma 5, every ideal of L is principal; therefore $L \cong I(L)$, and so L is Boolean. By exercise 6.21, L satisfies the Ascending Chain Condition; thus (exercise 6.22) every element of L other than the unit is contained in a dual atom. Since the complement of a dual atom is an atom, by taking complements we find that every nonzero element of L contains an atom.

If $p_0, p_1, \ldots, p_n, \ldots$ are distinct atoms in L, then the ascending chain $p_0, p_0 \vee p_1, \ldots, p_0 \vee p_1 \vee \cdots \vee p_n, \ldots$ does not terminate, and thus L has only finitely many atoms, $p_0, \ldots, p_{n-1}$. Let $a = p_0 \vee \cdots \vee p_{n-1}$. If $a' \neq 0$, then a' has to contain an atom, which is impossible. Therefore, $a' = 0$, $a = 1$, and $L \cong P(X)$, with $|X| = n$. ●

Exercises

1. Work out a direct proof of Theorem 2(i).
2. Work out a direct proof of Theorem 2(ii).

3. Let K be a five-element distributive lattice. Is there an identity $p = q$ such that $p = q$ holds in a lattice L iff L has no sublattice isomorphic to K?

4. Does the property stated in Lemma 5 characterize distributive lattices?

5. Let L be a distributive lattice with 0 and 1. Prove that the direct decompositions $L \cong L_0 \times L_1$ of L are in one-to-one correspondence with the complemented elements of L.

6. Prove that the complemented elements of a distributive lattice form a sublattice.

7. Let L be a distributive lattice with 0 and 1. Let $L \cong L_0 \times L_1 \cong K_0 \times K_1$. Show that there is a direct decomposition $L \cong A_0 \times A_1 \times A_2 \times A_3$ such that $A_0 \times A_1 \cong L_0$, $A_2 \times A_3 \cong L_1$, $A_0 \times A_2 \cong K_0$, $A_1 \times A_3 \cong K_1$.

8. Extend Theorem·9 to distributive lattices satisfying the Descending Chain Condition (see exercise 6.14).

9. Extend Corollary 10 to distributive lattices satisfying the Descending Chain Condition.

10. Can exercises 8 and 9 be further sharpened?

11. Let L be a distributive lattice with 0 and 1. Let C_0 and C_1 be finite chains in L. Show that there exist chains $D_0 \supseteq C_0$, $D_1 \supseteq C_1$, such that $|D_0| = |D_1|$.

12. Derive from exercise 11 the result that all maximal chains of a finite distributive lattice have the same length.

13. Find examples showing that exercise 11 is not valid if "finite" is omitted.

14. Prove the theorem "L is modular iff $I(L)$ is modular" by showing that "L has $\mathfrak{N}_5$ as a sublattice iff $I(L)$ has $\mathfrak{N}_5$ as a sublattice."

*15. Is the second statement of exercise 14 true for $\mathfrak{M}_5$ rather than for $\mathfrak{N}_5$?

16. Let L be a distributive lattice, $a,b,c \in L$, $a \le b$. Show that $[a, b]$ is Boolean iff $[a \wedge c, b \wedge c]$ and $[a \vee c, b \vee c]$ are Boolean.

17. For a poset P let $H_F(P)$ denote the lattice of all subsets of the form $(a_0] \cup (a_1] \cup \cdots \cup (a_{n-1}]$, where n is an arbitrary integer. Show that Theorem 9 holds for $H_F(P)$.

18. Prove that if P is a finite poset, then $H_F(P) = H(P)$.

19. Verify that if Theorem 9 holds for a lattice L, then $L \cong H_F(P)$ for some poset P.

20. Show that the Ascending Chain Condition implies the Descending Chain Condition for Boolean algebras.

21. Show that exercise 20 fails to hold for *generalized Boolean algebras* (that is, relatively complemented distributive lattices with 0).

22. Use exercise 20 to simplify the proof of Theorem 23.

23. Let L be a lattice, let P be a prime ideal of L, and let $a,b,c \in L$. Prove that if $a \vee (b \wedge c) \in P$, then $(a \vee b) \wedge (a \vee c) \in P$.

24. Using exercise 23, show that the lattice L is distributive iff, for all $x,y \in L$, $x < y$, there exists a prime ideal P with $x \in P$, $y \notin P$.

25. Verify the statement of exercise 24 using Theorem 6.

26. Verify the statement of exercise 24 using Theorem 1.
27. Let L be a distributive lattice. Then L is relatively complemented iff $\mathscr{P}(L)$ is unordered.
28. Prove Theorem 15 by well-ordering L, $L = \{a_\gamma \mid \gamma < \alpha\}$, and deciding one by one for each a_γ whether $a_\gamma \in P$ or $a'_\gamma \in P$.
29. Let L be a distributive lattice with 1. Show that every prime ideal P is contained in a *maximal prime ideal* Q (that is, $P \subseteq Q$, and for any prime ideal X of L, $Q \subseteq X$ implies that $Q = X$).
30. Let L be a distributive lattice with 0. Verify that every prime ideal P contains a *minimal prime ideal* Q (that is, $P \supseteq Q$, and for any prime ideal X of L, $Q \supseteq X$ implies that $Q = X$).
31. Find a distributive lattice L with no minimal and no maximal prime ideals.
32. Investigate the connections among the Ascending Chain Condition (and Descending Chain Condition) for a distributive lattice, for $I(L)$, and for $\mathscr{P}(L)$.
33. Let L be a distributive lattice with 0 and let $I \in I(L)$. Show that the pseudocomplement of I is $I^* = \{x \mid (x] \wedge I = (0]\}$.
34. Prove that $I = I^{**}$ for any $I \in I(L)$ of a distributive lattice with 0 iff L is a generalized Boolean lattice satisfying the Descending Chain Condition.
35. The congruence relations Θ and Φ *permute* if $a \equiv b(\Theta)$, $b \equiv c(\Phi)$ imply that $a \equiv d(\Phi)$, $d \equiv c(\Theta)$ for some d. Show that the congruences of a relatively complemented lattice permute.
36. Prove the converse of exercise 35 for distributive lattices.
37. Generalize Theorem 23 to distributive lattices without 0 and 1.
*38. Let L be a distributive lattice, let $a \in L$, let S be a sublattice of L, and let $a \notin S$. Show that there exists a prime ideal P and a prime dual ideal Q such that $a \notin P \cup Q \supseteq S$ provided that a is not the 0 or 1 of L (J. Hashimoto [1952]).
39. Let L be a relatively complemented distributive lattice. A sublattice K of L is *proper* if $K \neq L$. Show that every proper sublattice of L can be extended to a maximal proper sublattice of L (K. Takeuchi [1951]; see also J. Hashimoto [1952]; and G. Grätzer and E. T. Schmidt [1958d]).
40. Show that the statement of exercise 39 is not valid in general if L is not relatively complemented (K. Takeuchi [1951]).
41. Generalize Corollary 13 to infinite distributive lattices, claiming the unique irredundant representation of certain ideals as a meet of prime ideals.
42. If P is a prime ideal of L, then $(P]$ is a principal prime ideal of $I(L)$. Is the converse true?
43. Show that Corollary 17 characterizes distributivity.
44. A chain C in a poset P is called *maximal* if, for any chain D in P, $C \subseteq D$ implies that $C = D$. Using Zorn's Lemma, show that every chain is contained in a maximal chain.

8. Polynomials and Freeness

We can introduce an equivalence relation $\equiv$ for lattice polynomials: $p \equiv q$ iff p and q define the same functions in the class $\mathbf{D}$ of distributive lattices. More formally, if p and q are lattice polynomials (see Section 4), then $p \equiv q$ iff, for any distributive lattice L and $a_0, a_1, \ldots \in L$, we have $p(a_0, \ldots) = q(a_0, \ldots)$ (see Definitions 4.1 and 4.2). For an n-ary lattice polynomial p, let $[p]\mathbf{D}$ denote the set of all n-ary lattice polynomials q satisfying $p \equiv q$ and let $P_{\mathbf{D}}(n)$ denote the set of all these equivalence classes, that is, $P_{\mathbf{D}}(n) = \{[p]\mathbf{D} \mid p \in \mathbf{P}^{(n)}\}$. Observe that for $p, p_1, q, q_1 \in \mathbf{P}^{(n)}$ if $p \equiv p_1$ and $q \equiv q_1$, then $p \wedge q \equiv p_1 \wedge q_1$ and $p \vee q \equiv p_1 \vee q_1$. Thus $[p]\mathbf{D} \wedge [q]\mathbf{D} = [p \wedge q]\mathbf{D}$ and $[p]\mathbf{D} \vee [q]\mathbf{D} = [p \vee q]\mathbf{D}$ define $\wedge$ and $\vee$ on $P_{\mathbf{D}}(n)$. It is easily seen that $P_{\mathbf{D}}(n)$ is a distributive lattice and $[p]\mathbf{D} \leq [q]\mathbf{D}$ iff the inequality $p \leq q$ holds in the class $\mathbf{D}$.

To describe the structure of $P_{\mathbf{D}}(n)$, let $Q(n)$ denote the poset of all nonvoid subsets of $\{0, 1, \ldots, n - 1\}$.

THEOREM 1.
  (i) $P_{\mathbf{D}}(n)$ is isomorphic with $H(Q(n))$.
  (ii) $P_{\mathbf{D}}(n)$ is a free distributive lattice on n generators.
  (iii)[2] $2^n \leq |P_{\mathbf{D}}(n)| < 2^{2^n}$.
  (iv) A finitely generated distributive lattice is finite.

PROOF.
  (i) A lattice polynomial p is called a *meet-polynomial* if it is of the form $x_{i_0} \wedge \cdots \wedge x_{i_k}$. For $J \subseteq \{0, \ldots, n - 1\}$, $J \neq \varnothing$, set

$$p_J = \bigwedge (x_i \mid i \in J).$$

We claim that $[p_J]\mathbf{D} \leq [p_K]\mathbf{D}$ $(J, K \subseteq \{0, \ldots, n - 1\})$ iff $J \supseteq K$. The "if" part is obvious. Now let $J \not\supseteq K$; then there exists an $i \in K$ such that $i \notin J$. Consider the two-element chain $\mathfrak{C}_2$ and substitute $x_i = 0$ and $x_j = 1$ for $j \neq i$. Obviously, $p_J = 1$ and $p_K = 0$; thus the inequality $p_J \leq p_K$ fails in $\mathfrak{C}_2$ and therefore in $\mathbf{D}$.

Every lattice polynomial is equivalent to one of the form

$$\bigvee p_J;$$

---

[2] The problem of determining $|F_{\mathbf{D}}(n)|$ goes back to R. Dedekind [1900]; see R. Church [1940], E. N. Gilbert [1954], G. Hansel [1967], and D. Kleitman [1969].

because every x_i is of this form, the join of two such polynomials is of this form and the same holds for the meet in view of

$$\bigvee p_{J_i} \wedge \bigvee p_{K_j} \equiv \bigvee (p_{J_i} \wedge p_{K_j})$$

and

$$p_{J_i} \wedge p_{K_j} \equiv p_{J_i \cup K_j}.$$

Next we claim that $[p]\mathbf{D}$ is join-irreducible (in $P_{\mathbf{D}}(n)$) iff it is a $[p_J]\mathbf{D}$. Since every $[p]\mathbf{D} \in P_{\mathbf{D}}(n)$ is a join of some $[p_J]\mathbf{D}$, it suffices to prove that a $[p_J]\mathbf{D}$ is join-irreducible. Let

$$p_J \equiv \bigvee (p_{J_i} \mid i \in K, J_i \subseteq \{0, \ldots, n-1\});$$

$J \subseteq J_i$ follows from $[p_J]\mathbf{D} \geq [p_{J_i}]\mathbf{D}$. Now if $[p_J]\mathbf{D} > [p_{J_i}]\mathbf{D}$ for all i, then we have $J \subset J_i$. Choose $j_i \in J_i$, $j_i \notin J$ for all $i \in K$. In $\mathfrak{C}_2$ put $x_k = 0$ for all $k = j_i$ and $x_k = 1$ otherwise. Then $p_J = 1$, and $\bigvee (p_{J_i} \mid i \in K) = 0$, which is a contradiction.

To sum up, the join-irreducible elements form a poset isomorphic to $Q(n)$. A reference to Theorem 7.9 completes the proof of (i).   ▶

(ii) Let L be a distributive lattice, $a_0, \ldots, a_{n-1} \in L$. Then the map $x_i \to a_i$ can be extended to the homomorphism

$$[p]\mathbf{D} \to p(a_0, \ldots, a_{n-1}),$$

proving (ii).   ▶

(iii) This proof is obvious from (i).   ▶

(iv) This proof is obvious from (iii).   ●

Figure 5.6 is the diagram of $P_{\mathbf{D}}(3)$.

Boolean polynomials are defined exactly as lattice polynomials except that all five operations $\wedge$, $\vee$, $'$, 0, 1 are used in the formation of polynomials. A formal definition is the same as Definition 4.1 with two clauses added: If p is a Boolean polynomial, so is p'; 0 and 1 are Boolean polynomials. An n-ary Boolean polynomial p defines a function in n variables on any Boolean algebra B; $p(a_0, \ldots, a_{n-1})$ can be defined imitating Definition 4.2.

For the Boolean polynomials p and q, set $p \equiv q$ if, for any Boolean algebra B and $a_0, a_1, \ldots \in B$, we have $p(a_0, \ldots) = q(a_0, \ldots)$. Let $[p]\mathbf{B}$ denote the equivalence class containing p. Observe that $p \equiv q$ is equivalent to the identity $p = q$ holding in the class $\mathbf{B}$ of all Boolean algebras.

Let $P_\mathbf{B}(n)$ denote the set of all $[p]\mathbf{B}$, where p is an n-ary Boolean polynomial. It is easily seen that $[p]\mathbf{B} \wedge [q]\mathbf{B} = [p \wedge q]\mathbf{B}$, $[p]\mathbf{B} \vee [q]\mathbf{B} = [p \vee q]\mathbf{B}$, $([p]\mathbf{B})' = [p']\mathbf{B}$, $0 = [0]\mathbf{B}$, $1 = [1]\mathbf{B}$ defines the Boolean operations on $P_\mathbf{B}(n)$, and thus $P_\mathbf{B}(n)$ is a Boolean algebra.

THEOREM 2.
  (i) $P_\mathbf{B}(n)$ is isomorphic to $(\mathfrak{B}_2)^{2^n}$.
  (ii) $P_\mathbf{B}(n)$ is a free Boolean algebra on n generators.
  (iii) $|P_\mathbf{B}(n)| = 2^{2^n}$.
  (iv) A finitely generated Boolean algebra is finite.

PROOF. A Boolean polynomial is called *atomic* if it is of the form

$$x_0^{i_0} \wedge \cdots \wedge x_{n-1}^{i_{n-1}},$$

where $i_j = 0$ or 1, x^0 denotes x, x^1 denotes x'. For every $J \subseteq \{0, \ldots, n-1\}$ there is an atomic polynomial p_J for which $i_j = 0$ iff $j \in J$. The crucial statement is: $[p_{J_0}]\mathbf{B} \le [p_{J_1}]\mathbf{B}$ iff $J_0 = J_1$. Indeed, let $J_0 \ne J_1$. We make the following substitution in $\mathfrak{B}_2$: $x_i = 1$ if $i \in J_0$, $x_i = 0$ if $i \notin J_0$; this makes $p_{J_0} = 1$, $p_{J_1} = 0$, contradicting $[p_{J_0}]\mathbf{B} \le [p_{J_1}]\mathbf{B}$.

Let $B(n)$ be the set of all Boolean polynomials that are equivalent to one of the form $\bigvee (p_{J_i} \mid i \in K)$. Then $B(n)$ is closed under $\vee$ and $\wedge$, since

$$\bigvee p_{J_i} \wedge \bigvee p_{I_k} \equiv \bigvee (p_{J_i} \wedge p_{I_k}),$$

and $p_{J_i} \wedge p_{I_k} \equiv p_{J_k}$ if $J_i = I_k$, and $p_{J_i} \wedge p_{J_k} \equiv 0$ otherwise.

Now we prove by induction on n that $x_i, x_i' \in B(n)$ for $i < n$. For $n = 1$, x_0, x_0' are atomic polynomials, so $x_0, x_0' \in B(1)$. By induction, $x_0 \equiv \bigvee (p_{J_i} \mid i \in K)$, where p_{J_i} are atomic $(n-1)$-ary polynomials; then

$$x_0 \equiv x_0 \wedge (x_{n-1} \vee x_{n-1}') \equiv (x_0 \wedge x_{n-1}) \vee (x_0 \wedge x_{n-1}')$$

$$\equiv \bigvee (p_{J_i} \wedge x_{n-1} \mid i \in K) \vee \bigvee (p_{J_i} \wedge x_{n-1}' \mid i \in K),$$

and similarly for x_0'. Thus $x_0, x_0' \in B(n)$, and, by symmetry, $x_i, x_i' \in B(n)$ for all $i < n$. Since $(p_J)' \equiv \bigvee (x_i' \mid i \in J) \vee \bigvee (x_i \mid i \notin J)$, we conclude that $(p_J)' \in B(n)$; therefore $B(n)$ is closed under $'$. Thus $B(n)$ is closed under $\wedge, \vee, ', 0, 1$. Since $B(n)$ includes all x_i, $i < n$, $B(n)$ is the set of all n-ary Boolean polynomials.

Consequently, every $[p]\mathbf{B}$ is a join of atomic ones, the $[p]\mathbf{B}$ for p atomic polynomials are unordered and 2^n in number, implying (i) and (iii). The proof of (ii) is routine (same proof as that of Theorem 1(ii)), and (iv) follows trivially from (iii).   ●

We can use Theorems 1 and 2 to characterize free distributive lattices and free Boolean algebras, respectively.

THEOREM 3.   *Let L be a distributive lattice generated by $\{a_i \mid i \in I\}$. L is freely generated by the a_i iff the validity in L of a relation of the form*

$$\bigwedge (a_i \mid i \in I_0) \leq \bigvee (a_i \mid i \in I_1)$$

implies that $I_0 \cap I_1 \neq \varnothing$, for I_0, I_1 finite nonvoid subsets of I.

PROOF.   The "only if" part can be easily verified by using substitutions in $\mathfrak{C}_2$. For the converse, let F be the distributive lattice freely generated by x_i, $i \in I$, and let φ be the homomorphism of F into (in fact, onto) L satisfying $x_i\varphi = a_i$ for $i \in I$. It suffices to prove that for the lattice polynomials $p, q, p\varphi \leq q\varphi$ implies that $[p]\mathbf{D} \leq [q]\mathbf{D}$. (We think of the elements of F as equivalence classes of polynomials in the x_i, $i \in I$.)

Let

$$p \equiv \bigvee (\bigwedge (x_i \mid i \in I_j) \mid j \in J)$$

and

$$q \equiv \bigwedge (\bigvee (x_i \mid i \in K_t) \mid t \in T).$$

Then $p\varphi \leq q\varphi$ takes the form

$$\bigvee (\bigwedge (a_i \mid i \in I_j) \mid j \in J) \leq \bigwedge (\bigvee (a_i \mid i \in K_t) \mid t \in T),$$

which is equivalent to

$$\bigwedge (a_i \mid i \in I_j) \leq \bigvee (a_i \mid i \in K_t)$$

for all $j \in J$, $t \in T$. By assumption this implies that $I_j \cap K_t \neq \varnothing$ for all $j \in J$, $t \in T$; thus

$$\bigwedge (x_i \mid i \in J_j) \leq \bigvee (x_i \mid i \in K_t)$$

for all $j \in J$, $t \in T$; implying that $[p]\mathbf{D} \leq [q]\mathbf{D}$.   ●

THEOREM 4.   *Let B be a Boolean algebra generated by $\{a_i \mid i \in I\}$. Then B is freely generated by $\{a_i \mid i \in I\}$ iff, whenever I_0, I_1, J_0, J_1 are finite subsets* *of I with $I_0 \cup I_1 = J_0 \cup J_1$ and $I_0 \cap I_1 = \varnothing$, then*

$$\bigwedge (a_i \mid i \in I_0) \wedge \bigwedge (a_i' \mid i \in I_1) \le \bigwedge (a_i \mid i \in J_0) \wedge \bigwedge (a_i' \mid i \in J_1)$$

implies that $I_0 = J_0$ and $I_1 = J_1$.

PROOF. Again, the "only if" part is by substitution into $\mathfrak{B}_2$. On the other hand, clearly B is freely generated by $\{a_i \mid i \in I\}$ iff, for every finite subset $\bar{I}$ of I, the subalgebra $[\{a_i \mid i \in \bar{I}\}]$ is freely generated by $\{a_i \mid i \in \bar{I}\}$. By Theorem 2, the latter holds iff $[\{a_i \mid i \in \bar{I}\}]$ has $2^{2^{|\bar{I}|}}$ elements, which, in turn, is equivalent to $[\{a_i \mid i \in \bar{I}\}]$ having $2^{|\bar{I}|}$ atoms. Using the proof of Theorem 2 and the present hypothesis for $I_0 \cup I_1 = \bar{I}$, we can see that the elements of the form $\bigwedge (a_i \mid i \in I_0) \wedge \bigwedge (a_i' \mid i \in I_1)$, where $I_0 \cup I_1 = \bar{I}$ and $I_0 \cap I_1 = \varnothing$, are distinct atoms in $[\{a_i \mid i \in \bar{I}\}]$, thus completing the proof. ●

For a simpler variant of Theorem 4 see exercise 9.43.

Now we turn our attention to an important application of polynomials: finding homomorphisms.

THEOREM 5. *Let the Boolean algebra B be generated by the subalgebra B_1 and the element a. Let B_2 be a Boolean algebra and let φ be a homomorphism of B_1 into B_2. The extensions of φ to homomorphisms of B into B_2 are in one-to-one correspondence with the elements p of B_2 satisfying the following conditions:*

(i) *If $x \in B_1$, $x \le a$, then $x\varphi \le p$.*
(ii) *If $x \in B_1$, $x \ge a$, then $x\varphi \ge p$.*

To prepare for the proof of this theorem we verify a simple lemma, in which $+$ denotes the *symmetric difference*; that is,

$$x + y = (x' \wedge y) \vee (x \wedge y').$$

LEMMA 6. *Let the Boolean algebra B be generated by the subalgebra B_1 and the element a. Then every element x of B can be represented in the form*

$$x = (a \wedge x_0) \vee (a' \wedge x_1), \qquad x_0, x_1 \in B_1.$$

This representation is not unique. In fact,

$$(a \wedge x_0) \vee (a' \wedge x_1) = (a \wedge y_0) \vee (a' \wedge y_1), \qquad x_0, x_1, y_0, y_1 \in B_1$$

iff

$$a \le (x_0 + y_0)' \quad \text{and} \quad x_1 + y_1 \le a.$$

PROOF.   Let B_0 denote the set of all elements of B having such a representation. If $x \in B_1$, $x = (a \wedge x) \vee (a' \wedge x)$; thus $B_1 \subseteq B_0$. Also

$$a = (a \wedge 1) \vee (a' \wedge 0),$$

and so $a \in B_0$. Therefore, to show $B_0 = B$, it suffices to verify that B_0 is a subalgebra, which is left as an exercise. Now note that for $p, q \in B$, $p = q$ iff $p \wedge a = q \wedge a$ and $p \wedge a' = q \wedge a'$; thus $(a \wedge x_0) \vee (a' \wedge x_1) = (a \wedge y_0) \vee (a' \wedge y_1)$ iff $a \wedge x_0 = a \wedge y_0$ and $a' \wedge x_1 = a' \wedge y_1$. However, $a \wedge x_0 = a \wedge y_0$ is equivalent to $(a \wedge x_0) + (a \wedge y_0) = 0$; that is, $a \wedge (x_0 + y_0) = 0$ (see exercise 11), which is the same as $a \leq (x_0 + y_0)'$. Similarly, $a' \wedge x_1 = a' \wedge y$ iff $x_1 + y_1 \leq a$.   ●

PROOF OF THEOREM 5.   Let p be an element as specified and let the map $\psi: B \to B_2$ be defined as follows:

$$\psi: (a \wedge x_0) \vee (a' \wedge x_1) \to (p \wedge x_0\varphi) \vee (p' \wedge x_1\varphi).$$

By Lemma 6, ψ is defined on B. It is well defined, because if $(a \wedge x_0) \vee (a' \wedge x_1) = (a \wedge y_0) \vee (a' \wedge y_1)$, then $x_1 + y_1 \leq a \leq (x_0 + y_0)'$; thus $(x_1 + y_1)\varphi \leq p \leq (x_0 + y_0)'\varphi$, and therefore

$$x_1\varphi + y_1\varphi \leq p \leq (x_0\varphi + y_0\varphi)',$$

implying that $(p \wedge x_0\varphi) \vee (p' \wedge x_1\varphi) = (p \wedge y_0\varphi) \vee (p' \wedge y_1\varphi)$. It is routine to check that ψ is a homomorphism. Conversely, if ψ is an extension of φ to B, then ψ is uniquely determined by $p = a\psi$, and p satisfies (i) and (ii).   ●

COROLLARY 7.   *Let us assume the conditions of Theorem 5. In addition, let B_2 be complete. Set*

$$x_0 = \bigvee (x\varphi \mid x \in B_1, x \leq a) \quad and \quad x_1 = \bigwedge (x\varphi \mid x \in B_1, x \geq a).$$

Then the extensions of φ to B are in one-to-one correspondence with the elements of the interval $[x_0, x_1]$. In particular, there is always at least one such extension.

Exercises

1. Get lower and upper bounds for $|P_{\mathbf{D}}(n)|$ that are sharper than those given by Theorem 1(iii).

2. Work out the details of the last step in the proof of Theorem 1.

3. Let p_J be an atomic Boolean polynomial. Show that under the substitution $x_i = 1$ for $i \in J$ and $x_i = 0$ for $i \notin J$, we get $p_J = 1$ and $p_{J_0} = 0$ for all $J_0 \neq J$.

4. Let $f: \{0, 1\}^n \to \{0, 1\}$. Prove that there is an n-ary Boolean polynomial p that defines the function f on $\mathfrak{B}_2$ (see Definition 4.2). (Let $K = \{J \mid J \subseteq \{0, 1, \ldots, n - 1\}$ such that $f(x_0, \ldots, x_{n-1}) = 1$, where $x_i = 1$ if $i \in J$ and $x_i = 0$ if $i \notin J\}$; set $p = \bigvee (p_J \mid J \in K)$ and use exercise 3.)

5. Let B be a Boolean algebra. Prove that there is a one-to-one correspondence between n-ary Boolean polynomials over B and maps $\{0, 1\}^n \to \{0, 1\}$. In other words, all 0, 1 substitutions take 0 and 1 as values and determine p.

6. An *algebraic function* over a Boolean algebra B is built up inductively from the variables and the elements of B using $\wedge$, $\vee$, and $'$. Show that the n-ary algebraic functions are in one-to-one correspondence with maps $\{0, 1\}^n \to B$.

7. Let p be an n-ary algebraic function over the Boolean lattice B and Θ a congruence relation on B. Show that $a_i \equiv b_i(\Theta)$, $i < n$ implies that $p(a_0, \ldots, a_{n-1}) \equiv p(b_0, \ldots, b_{n-1})(\Theta)$.

8. Show that the property described in exercise 7 characterizes algebraic functions (G. Grätzer [1962]).

*9. Use the property described in exercise 7 to define *Boolean algebraic functions* over a distributive lattice. Show that for distributive lattices with 0 and 1, exercise 6 holds without any change (G. Grätzer [1964]).

10. Show that a free Boolean algebra on countably many generators has no atoms.

11. Show that $a \wedge (b + c) = (a \wedge b) + (a \wedge c)$ holds in any Boolean algebra.

12. Let B be the Boolean algebra freely generated by $\{x_i \mid i \in I\}$. Let L be the sublattice generated by $\{x_i \mid i \in I\}$. Prove that L is the free distributive lattice freely generated by $\{x_i \mid i \in I\}$.

13. Let L and L_1 be distributive lattices, let $L = [A]$, and let φ be a map of A into L_1. Show that there is a homomorphism of L into L_1 extending φ iff, for all pairs of n-ary lattice polynomials p and q and $a_0, \ldots, a_{n-1} \in A$, $p(a_0, \ldots, a_{n-1}) = q(a_0, \ldots, a_{n-1})$ implies that $p(a_0\varphi, \ldots, a_{n-1}\varphi) = q(a_0\varphi, \ldots, a_{n-1}\varphi)$.

14. State and prove exercise 13 for Boolean algebras.

15. Interpret Lemma 6 using exercise 14.

16. Extend the last statement of Corollary 7 to the case in which B_1 is generated by B and $a_0, \ldots, a_{n-1} \in B_1$, $n > 1$.

17. Let p and q be lattice polynomials. Since p and q can also be regarded as Boolean polynomials, $p \equiv q$ was defined in two ways in this section: with respect to **D** and with respect to **B**. Show that the two definitions are equivalent for lattice polynomials.

18. Define $\equiv$ for lattice polynomials with respect to a class **K** of lattices. Show that $P_{\mathbf{K}}(n) \in \mathbf{K}$ iff the free lattice over **K** with n generators exists, in which case $P_{\mathbf{K}}(n)$ is a free lattice with n generators.

9.   Congruence Relations

Let L be a lattice and let $H \subseteq L^2$. We denote by $\Theta(H)$ the smallest congruence relation such that $a \equiv b$ for all $\langle a, b \rangle \in H$.

LEMMA 1.   *For any $H \subseteq L^2$, $\Theta(H)$ exists.*

PROOF.   Let $\Phi = \bigwedge (\Theta \mid a \equiv b(\Theta)$ for all $\langle a, b \rangle \in H)$. Since in the lattice $C(L)$ the meet is intersection, it is obvious that $a \equiv b(\Phi)$ for all $\langle a, b \rangle \in H$; thus $\Phi = \Theta(H)$.   ●

We will use special notations in two cases: If $H = \{\langle a, b \rangle\}$, we write $\Theta(a, b)$ for $\Theta(H)$. If $H = I^2$, where I is an ideal, we write $\Theta[I]$ for $\Theta(H)$. The congruence relation $\Theta(a, b)$ is called *principal*; its importance is revealed by the following formula.

LEMMA 2. $\Theta(H) = \bigvee (\Theta(a, b) \mid \langle a, b \rangle \in H)$.

PROOF.   The proof is obvious.   ●

Note that $\Theta(a, b)$ is the smallest congruence relation under which $a \equiv b$, whereas $\Theta[I]$ is the smallest congruence relation under which I is contained in a single class.

In general lattices, not much can be said about $\Theta(H)$. In distributive lattices, the following description of $\Theta(a, b)$ is important (G. Grätzer and E. T. Schmidt [1958e]):

THEOREM 3.   *Let L be a distributive lattice, $a,b,x,y \in L$, and let $a \leq b$. Then $x \equiv y(\Theta(a, b))$ iff $x \wedge a = y \wedge a$ and $x \vee b = y \vee b$.*

REMARK.   This situation is illustrated in Figure 9.1.

PROOF.   Let Φ denote the binary relation under which $x \equiv y(\Phi)$ iff $x \wedge a = y \wedge a$ and $x \vee b = y \vee b$. Φ is obviously an equivalence relation. If $x \equiv y(\Phi)$ and $z \in L$, then $(x \vee z) \wedge a = (x \wedge a) \vee (z \wedge a) = (y \wedge a) \vee (z \wedge a) = (y \vee z) \wedge a$, and $(x \vee z) \vee b = z \vee (x \vee b) = z \vee (y \vee b) = (y \vee z) \vee b$; thus $x \vee z \equiv y \vee z(\Phi)$. Similarly, $x \wedge z \equiv y \wedge z(\Phi)$, and so Φ is a congruence relation. That $a \equiv b(\Phi)$ is obvious. Finally, let Θ be any congruence relation such that $a \equiv b(\Theta)$ and let $x \equiv y(\Phi)$. Therefore, $x \wedge a = y \wedge a$, $x \vee b = y \vee b$, $x \vee a \equiv x \vee b(\Theta)$,

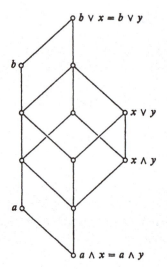

Figure 9.1

and $x \wedge b \equiv x \wedge a(\Theta)$. Then, computing modulo Θ, we obtain

$$x = x \vee (x \wedge a) = x \vee (y \wedge a)$$
$$= (x \vee y) \wedge (x \vee a) \equiv (x \vee y) \wedge (x \vee b)$$
$$= (x \vee y) \wedge (y \vee b) = y \vee (x \wedge b)$$
$$\equiv y \vee (x \wedge a) = y \vee (y \wedge a) = y,$$

that is, $x \equiv y(\Theta)$, proving that $\Phi \leq \Theta$. ●

EXPLANATION.   Since $a \equiv b$ implies that $(a \vee p) \wedge q \equiv (b \vee p) \wedge q$, we must have $x \equiv y(\Theta(a, b))$ if $x \wedge y = (a \vee p) \wedge q, x \vee y = (b \vee p) \wedge q$. It is easy to check that the x, y satisfying the conditions of Theorem 3 are exactly the same as those for which such p, q exist. Thus Theorem 3 can be interpreted as follows: We get all pairs x, y with $x \equiv y(\Theta(a, b))$ and $x \leq y$ by applying the substitution property "twice." No further application of the substitution property or transitivity is needed.

Some applications of Theorem 3 follow.

COROLLARY 4.   *Let I be an ideal of the distributive lattice L. Then $x \equiv y(\Theta[I])$ iff $x \vee y = (x \wedge y) \vee i$ for some $i \in I$. Therefore, I is a congruence class modulo $\Theta[I]$.*

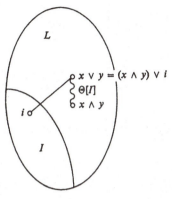

Figure 9.2

REMARK.   This situation is illustrated in Figure 9.2, in which the wavy line indicates congruence modulo $\Theta[I]$.

PROOF.   If $x \vee y = (x \wedge y) \vee i$, then $x \equiv y(\Theta(x \wedge y \wedge i, i))$, $x \wedge y \wedge i$, $i \in I$, and so $x \equiv y(\Theta[I])$. Conversely, $\Theta[I] = \bigvee (\Theta(u, v) \mid u,v \in I)$ by Lemma 2. However, $\Theta(u, v) \vee \Theta(u_1, v_1) \leq \Theta(u \wedge v \wedge u_1 \wedge v_1, u \vee v \vee u_1 \vee v_1)$; therefore $\Theta[I] = \bigcup(\Theta(u, v) \mid u,v \in I)$. If $x \equiv y(\Theta(u, v))$, $u,v \in I$, $u \leq v$, then $x \vee v = y \vee v$, and so $(x \wedge y) \vee [v \wedge (x \vee y)] = x \vee y$; thus the condition of Corollary 4 is satisfied with $i = v \wedge (x \vee y) \in I$. Finally, if $a \in I$, $a \equiv b(\Theta[I])$, then $a \vee b = (a \wedge b) \vee i$, $i \in I$, and so $a \vee b \in I$, and $b \in I$, showing that I is a full congruence class. ●

COROLLARY 5.   *Let L be a distributive lattice, $x,y,a,b \in L$, and let $x \leq y \leq a \leq b$ or $a \leq b \leq x \leq y$. Then $x \equiv y(\Theta(a, b))$ implies that $x = y$.*

A very important congruence relation has already been used in the proof of Lemma 3.5(ii): Given a prime ideal P of the lattice L, we can construct a congruence relation that has exactly two congruence classes, P and $L - P$. This statement can be generalized as follows: Let A be a set of prime ideals of a lattice L and let us call two elements x and y congruent modulo A iff, for every $P \in A$, either $x,y \in P$ or $x,y \in L - P$; this describes a congruence relation on L. For instance, if $A = \{P, Q, R\}$, $Q \subset P$, $R \subset P$, we get five congruence classes as shown in Figure 9.3; the quotient lattice is shown in Figure 5.8.

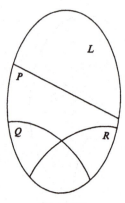

Figure 9.3

This principle will be used often. An interesting application is:

Theorem 6. *Let K be a sublattice of a distributive lattice L and let* Θ *be a congruence relation of K. Then* Θ *can be extended to L; that is, there exists a congruence relation* Φ *on L such that, for* $x, y \in K$, $x \equiv y(\Phi)$ *iff* $x \equiv y(\Theta)$.

Proof. Let φ be the natural homomorphism of K onto K/Θ, that is, $\varphi: x \to [x]\Theta$; then, for every prime ideal P of K/Θ, $P\varphi^{-1}$ is a prime ideal of K. Therefore, $(P\varphi^{-1}]$ is an ideal of L and $[K - P\varphi^{-1})$ is a dual ideal of L; thus by Theorem 7.15 we can choose a prime ideal P_1 of L such that $P_1 \supseteq P\varphi^{-1}$ and $P_1 \cap (K - P\varphi^{-1}) = \varnothing$. For every prime ideal P of K we choose such a prime ideal P_1 of L; let A denote the collection of all such prime ideals. Let Φ be the congruence relation associated with A as previously described. Now for $x, y \in K$ the condition $x \equiv y(\Theta)$ is equivalent to $x\varphi = y\varphi$, and so for every $P_1 \in A$ either $x, y \in P_1$ or $x, y \notin P_1$; thus $x \equiv y(\Phi)$. Conversely, if $x \equiv y(\Phi)$, then for every $P_1 \in A$ either $x, y \in P_1$ or $x, y \notin P_1$, and so either $x\varphi, y\varphi \in P$ or $x\varphi, y\varphi \notin P$. Since every pair of distinct elements of K/Θ is separated by a prime ideal (Corollary 7.17), we conclude that $x\varphi = y\varphi$ and thus $x \equiv y(\Theta)$. ●

It is well known that in rings, ideals are in a one-to-one correspondence with congruence relations. In one class of lattices the situation is exactly the same as in the class of rings.

Theorem 7. *Let L be a Boolean lattice. Then*

$$\Theta \to [0]\Theta$$

is a one-to-one correspondence between congruence relations and ideals of L.

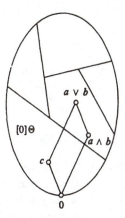

Figure 9.4

PROOF.   By Corollary 4 the map is onto; therefore, we have only to prove that it is one-to-one, that is, that $[0]\Theta$ determines Θ. This fact, however, is obvious, since $a \equiv b(\Theta)$ iff $a \wedge b \equiv a \vee b(\Theta)$, which in turn is equivalent to $c \equiv 0(\Theta)$, where c is the relative complement of $a \wedge b$ in $[0, a \vee b]$ (see Figure 9.4). Thus $a \equiv b(\Theta)$ iff $c \in [0]\Theta$.   ●

   This proof does not make full use of the hypothesis that L is a complemented distributive lattice. In fact, all we need to make the proof work is that L has a zero and is relatively complemented. Such a distributive lattice is called a *generalized Boolean lattice*.

THEOREM 8   (J. Hashimoto [1952]).   *Let L be a lattice. There is a one-to-one correspondence between ideals and congruence relations of L under which the ideal corresponding to a congruence relation, Θ, is a whole congruence class under Θ iff L is a generalized Boolean algebra.*

PROOF   (G. Grätzer and E. T. Schmidt [1958e]).   The proof of the "if" part is the proof of Theorem 7. We proceed with the "only if" part. The ideal corresponding to ω has to be (0], and thus L has a 0. If L has $\mathfrak{M}_5$ as a sublattice, then (using the notation of Figure 7.1) $(a]$ cannot be a congruence class, because $a \equiv o$ implies that $i = a \vee c \equiv o \vee c = c$, and $b = b \wedge i \equiv b \wedge c = o$. But $o \in (a]$, and thus any congruence class containing $(a]$ contains b, and $b \notin (a]$. Similarly, if L contains $\mathfrak{M}_5$, and a congruence class contains $(b]$, then $b \equiv o$; thus $i = b \vee c \equiv o \vee c = c$, and so $a = a \wedge i \equiv a \wedge c = o$. Therefore, this congruence class has to contain a, and $a \notin (b]$. Thus, by Theorem 7.1, L is distributive. Let $a < b$,

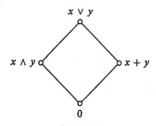

Figure 9.5

$I = [0]\Theta(a, b)$. By Corollary 4, $\Theta[I]$ is also a congruence relation of L having I as a whole congruence class; consequently, $\Theta[I] = \Theta(a, b)$, and so $a \equiv b(\Theta[I])$. Thus, again by Corollary 4, $b = a \vee i$ for some $i \in I$, and $i \equiv 0(\Theta(a, b))$. The latter is equivalent to $i \vee b = 0 \vee b$ and $i \wedge a = 0 \wedge a$. Thus $a \vee i = b$ and $a \wedge i = 0$, and so i is the relative complement of a in $[0, b]$. ●

It is no coincidence that, in the class of generalized Boolean lattices, congruences and ideals behave as they do in rings. Indeed, generalized Boolean lattices are rings in disguise.

THEOREM 9 (M. H. Stone [1936]).

(i) *Let* $\mathfrak{B} = \langle B; \wedge, \vee \rangle$ *be a generalized Boolean lattice. Define the binary operations* $\cdot$ *and* $+$ *on B by setting*

$$x \cdot y = x \wedge y$$

and by defining $x + y$ *as the relative complement of* $x \wedge y$ *in* $[0, x \vee y]$ *(see Figure 9.5). Then* $\mathfrak{B}^R = \langle B; +, \cdot \rangle$ *is a* Boolean ring— *that is, an (associative) ring satisfying* $x^2 = x$, *for all* $x \in B$ *(and, consequently, satisfying* $xy = yx$ *and* $x + x = 0$ *for* $x, y \in B$).

(ii) *Let* $\mathfrak{B} = \langle B; +, \cdot \rangle$ *be a Boolean ring. Define the binary operations* $\wedge$ *and* $\vee$ *on B by*

$$x \wedge y = x \cdot y$$

and

$$x \vee y = x + y + x \cdot y.$$

Then $\mathfrak{B}^L = \langle B; \wedge, \vee \rangle$ *is a generalized Boolean lattice.*

(iii) *Let* $\mathfrak{B}$ *be a generalized Boolean lattice. Then* $(\mathfrak{B}^R)^L = \mathfrak{B}$.

(iv) *Let* $\mathfrak{B}$ *be a Boolean ring. Then* $(\mathfrak{B}^L)^R = \mathfrak{B}$.

The proof of this theorem is purely computational. Some steps will be given in the exercises. The correspondence between Boolean rings and generalized Boolean lattices preserves many algebraic properties.

THEOREM 10.   *Let $\mathfrak{B}_0$ and $\mathfrak{B}_1$ be generalized Boolean lattices.*

 (i) *Let $I \subseteq B_0$. Then I is an ideal of $\mathfrak{B}_0$ iff I is an ideal of $\mathfrak{B}_0^R$.*

 (ii) *Let $\varphi\colon B_0 \to B_1$. Then φ is a $\{0\}$-homomorphism of $\mathfrak{B}_0$ into $\mathfrak{B}_1$ iff φ is a homomorphism of $\mathfrak{B}_0^R$ into $\mathfrak{B}_1^R$.*

 (iii) *$\mathfrak{B}_0$ is a $\{0\}$-sublattice of $\mathfrak{B}_1$ iff $\mathfrak{B}_0^R$ is a subring of $\mathfrak{B}_1^R$.*

The proof is again left to the reader.

Congruence relations on an arbitrary lattice have an interesting connection with distributive lattices:

THEOREM 11   (N. Funayama and T. Nakayama [1942]).   *Let L be an arbitrary lattice. Then $C(L)$, the lattice of all congruence relations of L, is distributive.*

PROOF.   Let $\Theta, \Phi, \Psi \in C(L)$. Since $\Theta \wedge (\Phi \vee \Psi) \geq (\Theta \wedge \Phi) \vee (\Theta \wedge \Psi)$, it suffices to prove that $a \equiv b(\Theta \wedge (\Phi \vee \Psi))$ implies that $a \equiv b((\Theta \wedge \Phi) \vee (\Theta \wedge \Psi))$. So let $a \equiv b(\Theta \wedge (\Phi \vee \Psi))$; that is, $a \equiv b(\Theta)$ and $a \equiv b(\Phi \vee \Psi)$. By Theorem 3.9, there exists a sequence $a \wedge b = z_0 \leq \cdots \leq z_n = a \vee b$ such that $z_i \equiv z_{i+1}(\Phi)$ or $z_i \equiv z_{i+1}(\Psi)$ for every $0 \leq i < n$. Since $a \equiv b(\Theta)$, we also have $a \wedge b \equiv a \vee b(\Theta)$, and so $z_i \equiv z_{i+1}(\Theta)$ for every $0 \leq i < n$. Thus for every $0 \leq i < n$, $z_i \equiv z_{i+1}(\Theta \wedge \Phi)$ or $z_i \equiv z_{i+1}(\Theta \wedge \Psi)$, implying that $a \equiv b((\Theta \wedge \Phi) \vee (\Theta \wedge \Psi))$. ●

Another property of congruence lattices is given in the following definition.

DEFINITION 12.

 (i) *Let L be a complete lattice and let a be an element of L. Then a is called* compact *if $a \leq \bigvee X$ for some $X \subseteq L$ implies that $a \leq \bigvee X_1$ for some finite $X_1 \subseteq X$.*

 (ii) *A complete lattice is called* algebraic *if every element is the join of compact elements.*

In the literature, algebraic lattices are also called *compactly generated lattices.*

Just as for lattices, a nonvoid subset I of a join-semilattice F is an ideal if, for $a,b \in F$, we have $a \vee b \in I$ iff a and $b \in I$. Again, $I(F)$ is the poset (not necessarily a lattice) of all ideals of F partially ordered under set inclusion. If F has a zero, then $I(F)$ is a lattice.

Using $I(F)$, we give a useful characterization of algebraic lattices:

THEOREM 13.   *A lattice L is algebraic iff it is isomorphic to the lattice of all ideals of a join-semilattice with 0.*

PROOF.   Let F be a join-semilattice with 0; we want to prove that $I(F)$ is algebraic. We know that $I(F)$ is complete. We claim that for $a \in F$, $(a]$ is a compact element of $I(F)$. Let $X \subseteq I(F)$ and let

$$(a] \subseteq \bigvee (I \mid I \in X).$$

Just as in the proof of Corollary 3.2,

$$\bigvee (I \mid I \in X) = \{x \mid x \leq t_0 \vee \cdots \vee t_{n-1}, t_i \in I_i, I_i \in X\}.$$

Therefore, $a \leq t_0 \vee \cdots \vee t_{n-1}$, $t_i \in I_i$, $I_i \in X$. Thus with

$$X_1 = \{I_0, \ldots, I_{n-1}\},$$

$$(a] \subseteq \bigvee (I \mid I \in X_1).$$

Since for any $I \in I(F)$ we have

$$I = \bigvee ((a] \mid a \in I),$$

we see that $I(F)$ is algebraic.

Now let L be an algebraic lattice and let F be the set of compact elements of L. Obviously, $0 \in F$. Let $a,b \in F$, $a \vee b \leq \bigvee X$, $X \subseteq L$. Then $a \leq a \vee b \leq \bigvee X$, and so $a \leq \bigvee X_0$, for some finite $X_0 \subseteq X$. Similarly, $b \leq \bigvee X_1$, for some finite $X_1 \subseteq X$. Thus $a \vee b \leq \bigvee (X_0 \cup X_1)$, and $X_0 \cup X_1$ is a finite subset of X. So $a \vee b \in F$.

Therefore, $\langle F; \vee \rangle$ is a join-semilattice with 0. Consider the map:

$$\varphi: a \to \{x \mid x \in F, x \leq a\}, \qquad a \in L.$$

Obviously, φ maps L into $I(F)$. By the definition of an algebraic lattice,

$$a = \bigvee a\varphi,$$

and thus φ is one-to-one. To prove that φ is onto, let $I \in I(F)$, $a = \bigvee I$. Then $a\varphi \supseteq I$. Let $x \in a\varphi$. Then $x \leq \bigvee I$, so that by the compactness of x,

$x \leq \bigvee I_1$ for some finite $I_1 \subseteq I$. Therefore $x \in I$, proving that $a\varphi \subseteq I$. Consequently, $a\varphi = I$, and so φ is onto. Thus φ is an isomorphism. ●
Now we connect the foregoing with congruence lattices.

LEMMA 14.   *Every principal congruence relation is compact.*

PROOF.   Let L be a lattice, $a,b \in L$, $X \subseteq C(L)$,

$$\Theta(a, b) \leq \bigvee (\Theta \mid \Theta \in X).$$

Then $a \equiv b(\bigvee (\Theta \mid \Theta \in X))$, and thus (just as in Theorem 3.9) there exists a sequence $a = x_0, x_1, \ldots, x_{n-1} = b$, in which $x_i \equiv x_{i+1}(\Theta_i)$ for some $\Theta_i \in X$. Therefore, $a \equiv b(\bigvee (\Theta \mid \Theta \in X_0))$, where $X_0 = \{\Theta_0, \ldots, \Theta_{n-1}\}$, and so $\Theta(a, b) \leq \bigvee (\Theta \mid \Theta \in X_0)$, where X_0 is a finite subset of X. ●

THEOREM 15.   *Let L be an arbitrary lattice. Then $C(L)$ is an algebraic lattice.*

PROOF.   For every $\Theta \in C(L)$,

$$\Theta = \bigvee (\Theta(a, b) \mid a \equiv b(\Theta)).$$

Consequently, this theorem follows from Lemma 14 and Corollary 3.15. ●
Combining Theorems 11 and 15 we get:

COROLLARY 16.   *Let L be an arbitrary lattice. Then $C(L)$ is a distributive algebraic lattice.*

The converse of Corollary 16 is a long-standing conjecture of lattice theory. We shall verify the conjecture in the finite case. This was first established by R. P. Dilworth. The present proof combines a construction of G. Grätzer and E. T. Schmidt [1962] (Lemma 18) with a result of G. Grätzer and H. Lakser [1968] (Lemma 19).

THEOREM 17.[3]   *Let K be a finite distributive lattice. Then there exists a finite lattice L such that K is isomorphic to $C(L)$.*

---

[3] The reader is advised to page over the proof of Theorem 17 at the first reading of this book.

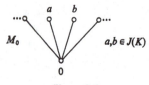

Figure 9.6

PROOF. The proof of Theorem 17 is immediate from Lemmas 18 and 19. We take $L = I(M)$ where M is given in Lemma 18. By Lemma 19, $C(L) \cong C(M)$; by Lemma 18, $C(M) \cong K$. Since M is finite, so is L. ●

Let M be a finite poset such that $\inf\{a, b\}$ exists in M for any $a, b \in M$. We define in M: $a \wedge b = \inf\{a, b\}$ and $a \vee b = \sup\{a, b\}$ whenever it exists. This makes M into a partial lattice. An equivalence relation Θ on M is a *congruence relation* if $a_0 \equiv b_0(\Theta)$ and $a_1 \equiv b_1(\Theta)$ imply that $a_0 \wedge a_1 \equiv b_0 \wedge b_1(\Theta)$ and that $a_0 \vee a_1 \equiv b_0 \vee b_1(\Theta)$ whenever $a_0 \vee a_1$ and $b_0 \vee b_1$ exist. Then the set $C(M)$ of all congruence relations is again a lattice.

LEMMA 18. *Let K be a finite distributive lattice. Then there exists a finite poset M such that $\inf\{a, b\}$ exists for any $a, b \in M$ and $C(M)$ is isomorphic to K.*

PROOF. Take the set $M_0 = J(K) \cup \{0\}$ and make it a meet-semilattice by defining $\inf\{a, b\} = 0$ if $a \neq b$ ($J(K)$ is the set of nonzero join-irreducible elements of K; see Section 7), as illustrated in Figure 9.6. Note that $a \equiv b(\Theta)$ and $a \neq b$ imply in M_0 that $a \equiv 0(\Theta)$ and $b \equiv 0(\Theta)$; therefore, congruence relations of M_0 are in one-to-one correspondence with subsets of $J(K)$. Thus $C(M_0)$ is a Boolean lattice whose atoms are associated with elements of $J(K)$; the congruence Φ_a associated with $a \in J(K)$ is: $x \equiv y(\Phi_a)$ if $\{x, y\} = \{a, 0\}$, and if $\{x, y\} \neq \{a, 0\}$, then $x \equiv y(\Phi_a)$ implies that $x = y$.

If $J(K)$ is unordered in K, then we are ready. However, if, say, $a, b \in J(K)$, $a > b$ in K, then we must have $\Phi_a > \Phi_b$. The simplest way to make this happen is to use the lattice $M(a, b)$ of Figure 9.7. Note that $M(a, b)$ has three congruence relations, namely, ω, ι, and Θ, where Θ is the congruence relation with congruence classes $\{0, b_1, b_2, b\}$, $\{a_1, a(b)\}$. Thus $\Theta(a_1, 0) = \iota$. In other words, $a_1 \equiv 0$ "implies" that $b_1 \equiv 0$, but $b_1 \equiv 0$ does not "imply" that $a_1 \equiv 0$.

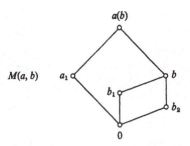

Figure 9.7

We construct M by "inserting" $M(a, b)$ in M_0 whenever $a > b$ in $J(K)$. Figure 9.8 gives M for the three-element chain.

More precisely, M consists of four kinds of elements: (i) 0; (ii) all maximal join-irreducible elements of K (that is, all $a \in J(K)$ such that there is no $x \in J(K)$ with $a < x$ in K); (iii) for any nonmaximal join-irreducible element a of K, three elements: a, a_1, a_2; (iv) for each pair $a,b \in J(K)$ with $a > b$, a new element, $a(b)$. To simplify the notation, for each maximal join-irreducible element a, we write $a = a_1 = a_2$. For $a,b \in J(K)$ with $a > b$, we set $M(a, b) = \{0, a_1, b, b_1, b_2, a(b)\}$.

Observe that $M(a, b) \cap M(c, d) = M(a, b)$ if $a = c$ and $b = d$; $M(a, b) \cap M(c, d) = \{0, b, b_1, b_2\}$ if $a \neq c$ and $b = d$; $M(a, b) \cap M(c, d) = \{0, a_1\}$ if $a = c$ and $b \neq d$; $M(a, b) \cap M(c, d) = \{0, b_1\}$ if $b = c$; otherwise, $M(a, b) \cap M(c, d) = \{0\}$.

For $x,y \in M$, let us define $x \leq y$ to mean that for some $a,b \in J(K)$ with $a > b$, we have $x,y \in M(a, b)$ and $x \leq y$ in the lattice $M(a, b)$ as illustrated in Figure 9.7. It is easily seen that $x \leq y$ does not depend on the choice of a, b and that $\leq$ is a partial ordering relation. Since, under this

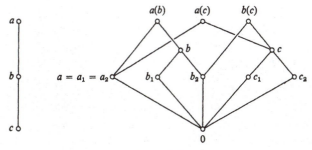

Figure 9.8

partial ordering, all $M(a, b)$ and $M(a, b) \cap M(c, d)$ are lattices and $x,y \in M$, $x \in M(a, b)$, and $y \leq x$ imply that $y \in M(a, b)$, we conclude that inf $\{u, v\}$ exist for all $u,v \in M$.

Now we describe $C(M)$. Let $H \in H(J(K))$ (notation of Definition 7.8). We define a binary relation Θ_H on M:

$x \equiv y(\Theta_H)$ if

either $x,y \in \bigcup (M(a, b) \mid a,b \in H, a > b) \cup \bigcup (\{0, a_1, a_2, a\} \mid a \in H)$,

or $x,y \in \{a_1, a(b), a(c)\}$, where $a > b$, $a > c$, $b,c \in H$, or $x = y$.

In other words, $[0]\Theta_H$ contains all a_1, a_2, a with $a \in H$; and if $a > b$, $a,b \in H$, then it also contains $a(b)$. Outside this class the only nontrivial congruence is $a(b) \equiv a_1 \equiv a(c)$, where $a \notin H$, and $b,c \in H$, $a > b$, $a > c$.

Θ_H is obviously an equivalence relation. The fact that Θ_H restricted to any $M(a, b)$ is a congruence relation easily implies that Θ_H is a congruence relation. Given a Θ_H we get

$$H = \{a \mid a_1 \equiv 0(\Theta_H)\};$$

thus the map

$$\varphi: H \to \Theta_H$$

is a one-to-one order preserving map of $H(J(K))$ into $C(M)$. To show that φ is an isomorphism, we have to show that φ is onto. So let Θ be a congruence relation of M, and

$$H = \{a \mid a_1 \equiv 0(\Theta)\}.$$

Since in $M(a, b)$ every congruence Θ is determined by the atoms in $[0]\Theta$, the same holds in M. Therefore, $\Theta = \Theta_H$. Thus $H(J(K)) \cong C(M)$. By Theorem 7.9, $K \cong H(J(K))$, and so $K \cong C(M)$. ●

LEMMA 19. *Let M be a finite poset with the property that inf $\{a, b\}$ exists for any $a,b \in M$. Then for every congruence relation Θ there exists exactly one congruence relation $\overline{\Theta}$ of $I(M)$ such that for $a,b \in M$, $(a] \equiv (b](\overline{\Theta})$ iff $a \equiv b(\Theta)$.*

PROOF. Since arbitrary meets exist in M, for every element $m \in M$, $(m]$ is a (finite) lattice, and so if $\{x, y\}$ has an upper bound, then $x \vee y$ exists.

Let Θ be a congruence relation of M. For $X \subseteq M$ set $[X]\Theta = \bigcup ([x]\Theta \mid x \in X)$; that is, $[X]\Theta = \{y \mid x \equiv y(\Theta)$ for some $x \in X\}$. If $I, J \in I(M)$, define $I \equiv J(\overline{\Theta})$ iff $[I]\Theta = [J]\Theta$. Obviously, $\overline{\Theta}$ is an equivalence relation. Let $I \equiv J(\overline{\Theta})$, $N \in I(M)$, and $x \in I \cap N$. Then $x \equiv y(\Theta)$ for some $y \in J$, and so $x \equiv x \wedge y(\Theta)$ and $x \wedge y \in J \cap N$. This shows that $[I \cap N]\Theta \subseteq [J \cap N]\Theta$. Similarly, $[J \cap N]\Theta \subseteq [I \cap N]\Theta$, so $I \cap N \equiv J \cap N(\overline{\Theta})$. To show that $I \vee N \equiv J \vee N(\overline{\Theta})$, recall the description of $I \vee N$ given in exercise 5.22: Set $A_0 = I \cup N$, and for $0 < n < \omega$, $A_n = \{x \mid x \leq t_0 \vee t_1, t_0, t_1 \in A_{n-1}\}$; then $I \vee N = \bigcup (A_n \mid n < \omega)$. We prove by induction on n that $A_n \subseteq [J \vee N]\Theta$. For $n = 0$, $A_0 = I \cup N \subseteq [J]\Theta \cup N \subseteq [J \vee N]\Theta$. Suppose that $A_{n-1} \subseteq [J \vee N]\Theta$ and let $x \in A_n$. Then $x \leq t_0 \vee t_1$ for some $t_0, t_1 \in A_{n-1}$. Thus $t_0 \equiv u_0(\Theta)$ and $t_1 \equiv u_1(\Theta)$ for some $u_0, u_1 \in J \vee N$, and so $t_0 \equiv t_0 \wedge u_0(\Theta)$ and $t_1 \equiv t_1 \wedge u_1(\Theta)$. Observe that $t_0 \vee t_1$ is an upper bound for $\{t_0 \wedge u_0, t_1 \wedge u_1\}$; consequently, $(t_0 \wedge u_0) \vee (t_1 \wedge u_1)$ exists. Therefore, $t_0 \vee t_1 \equiv (t_0 \wedge u_0) \vee (t_1 \wedge u_1)(\Theta)$. Finally, $x = x \wedge (t_0 \vee t_1) \equiv x \wedge ((t_0 \wedge u_0) \vee (t_1 \wedge u_1))(\Theta)$ and $x \wedge ((t_0 \wedge u_0) \vee (t_1 \wedge u_1)) \in J \vee N$.

Thus $x \in [J \vee N]\Theta$. Since $I \vee N = \bigcup (A_n \mid n < \omega)$, we conclude that $I \vee N \subseteq [J \vee N]\Theta$. Similarly, $J \vee N \subseteq [I \vee N]\Theta$, proving that $I \vee N \equiv J \vee N(\overline{\Theta})$. This completes the verification that $\overline{\Theta}$ is a congruence relation of $I(M)$.

If $a \equiv b(\Theta)$ and $x \in (a]$, then $x \equiv x \wedge b(\Theta)$. Thus $(a] \subseteq [(b]]\Theta$. Similarly, $(b] \subseteq [(a]]\Theta$, and so $(a] \equiv (b](\overline{\Theta})$. Conversely, if $(a] \equiv (b](\overline{\Theta})$, then $a \equiv b_1(\Theta)$ and $a_1 \equiv b(\Theta)$ for some $a_1 \leq a$ and $b_1 \leq b$. Forming the join of the two congruences, we get $a \equiv b(\Theta)$. Thus $\overline{\Theta}$ has all the properties required by Lemma 19.

To show the uniqueness, let Φ be a congruence relation of $I(M)$ satisfying $(a] \equiv (b](\Phi)$ iff $a \equiv b(\Theta)$. Let $I, J \in I(M)$, $I \equiv J(\Phi)$, and $x \in I$. Then $(x] \cap I \equiv (x] \cap J(\Phi)$, $(x] \cap I = (x]$, and $(x] \cap J = (y]$ for some $y \in J$. Thus $(x] \equiv (y](\overline{\Phi})$, and so $x \equiv y(\Theta)$, proving that $I \subseteq [J]\Theta$. Similarly, $J \subseteq [I]\Theta$, and so $I \equiv J(\overline{\Theta})$. Conversely, if $I \equiv J(\overline{\Theta})$, then take all congruences of the form $x \equiv y(\Theta)$, $x \in I$, $y \in J$. By our assumption regarding Φ, $(x] \equiv (y](\Phi)$, and by our definition of $\overline{\Theta}$, the join of all these congruences yields $I \equiv J(\Phi)$. Thus $\Phi = \overline{\Theta}$.   ●

More general results along these lines can be found in G. Grätzer and E. T. Schmidt [1962], G. Grätzer and H. Lakser [1968], and E. T. Schmidt [1968] and [1969].

Exercises

1. Let L be a distributive lattice and let $x,y,a,b \in L$. Prove that if $a \leq b$ and $x \leq y$, then $x \wedge a = y \wedge a$ and $x \vee b = y \vee b$ is equivalent to $(a \vee p) \wedge q = x$ and $(b \vee p) \wedge q = y$ for some p, q in L.

2. Verify Corollary 4 directly.

3. Let K be a sublattice of the distributive lattice L and let P be a prime ideal of K. Prove that there exists a prime ideal Q of L with $Q \cap K = P$.

4. Prove that Corollary 5 characterizes the distributivity of L.

5. Show that Theorem 6 characterizes the distributivity of L.

6. What is the smallest n such that Theorem 6 for $|K| \leq n$ characterizes the distributivity of L?

7. Let L be a *sectionally complemented lattice*; that is, L has a 0 and all intervals $[0, a]$ are complemented. Prove that $\Theta \rightarrow [0]\Theta$ is a one-to-one correspondence between congruences and *certain* ideals of L.

*8. Show that the "certain ideals" that appear in exercise 7 form a sublattice of $I(L)$.

9. Prove that every (principal) ideal of L is of the form $[0]\Theta$ iff L is distributive.

10. Let L be a distributive lattice and let I be an ideal of L. Define a binary relation on $L: x \equiv y(\Phi(I))$ iff there is no $a \in L$ with $a \leq x \vee y$, $x \wedge y \wedge a \in I$, $a \notin I$. Prove that $\Phi(I)$ is the largest congruence relation of L under which I is a whole congruence class.

11. Let L be a distributive lattice with 0. Prove that there is a one-to-one correspondence between ideals and congruence relations (in the sense of Theorem 8) iff $\Theta[I] = \Phi(I)$ for all $I \in I(L)$.

12. Prove Theorem 8 using exercises 10 and 11 (G. Ja. Areškin [1953a]).

*13. Let L be a lattice and let a be an element of L. Show that every convex sublattice of L containing a is a congruence class under exactly one congruence relation iff L is distributive and all the intervals $[b, a]$ ($b \in L$, $b \leq a$) and $[a, c]$ ($c \in L$, $a \leq c$) are complemented (G. Grätzer and E. T. Schmidt [1958e]).

14. Derive Theorem 8 (and, also, a variant of Theorem 8) by taking $a = 0$ (arbitrary $a \in L$) in exercise 13.

15. Let L be a relatively complemented lattice, $I, J \in I(L)$, $I \subseteq J$. Prove that if I is an intersection of prime ideals, then so is J (J. Hashimoto [1952]).

16. Use exercises 14 and 15 to get the following theorem: *Let L be a relatively complemented lattice. Then L is distributive iff, for some element a of L, $(a]$ is an intersection of prime ideals and $[a)$ is an intersection of prime dual ideals* (J. Hashimoto [1952]).

17. Prove that the verification of Theorem 9(i) can be reduced to the Boolean lattice case and that in this case $x + y = (x \wedge y') \vee (x' \wedge y)$.

18. Let B be a Boolean lattice. Verify that $x + y = (x \wedge y) \vee (x' \wedge y')$.

19. Let B be a Boolean lattice. Verify that $(x + y) + z = (x \wedge y \wedge z) \vee (x' \wedge y' \wedge z) \vee (x \wedge y' \wedge z') \vee (x' \wedge y \wedge z')$ and conclude that $+$ is associative.

20. Prove that $x(y + z) = xy + xz$.
21. Prove Theorem 9(i).
22. Prove Theorem 9(ii).
23. Let $\mathfrak{B}$ be a generalized Boolean lattice. Observe that for $x,y \in B$, $x \wedge y$ is the same in $\mathfrak{B}$ as in $(\mathfrak{B}^R)^L$ (namely, $x \cdot y$); conclude that $\mathfrak{B} = (\mathfrak{B}^R)^L$.
24. Verify Theorem 9(iv).
25. Verify Theorem 10.
26. Show that, using the concept of a distributive semilattice (see Section 11), Corollary 16 can be reformulated as follows: Let L be an arbitrary lattice. Prove that there exists a distributive join-semilattice F with 0 such that $C(L)$ is isomorphic to $I(F)$.
27. Characterize the lattice of all ideals of a lattice using the concept of an algebraic lattice.
28. Characterize the lattice of all ideals of a Boolean lattice.
29. Let M be a poset such that $a \wedge b$ exists for all $a,b \in M$. Let Θ and Φ be congruences of M. Find a result for M analogous with Lemma 3.8.
30. Let M be as in exercise 29.   Generalize Theorem 3.9 to M.
31. Let M be as in exercise 29. Is $C(M)$ necessarily distributive?
32. Let M be as in exercise 29. Assume that M satisfies the following conditions:
    (i) There exists $H \subseteq M$ such that for $h \in H$, $(h]$ is a lattice.
    (ii) For each ideal I that belongs to the sublattice $I_P(M)$ of $I(M)$ generated by the principal ideals, there exist finite $\{h_1, \ldots, h_n\} \subseteq H$, $\{i_1, \ldots, i_n\} \subseteq I$ with $i_j \leq h_j$ and $I = (i_1] \cup \cdots \cup (i_n]$.
    Under these conditions, prove Lemma 19 for M, replacing $I(M)$ with $I_P(M)$ (G. Grätzer and H. Lakser [1968]).

Exercises 33–36 are from G. Grätzer and E. T. Schmidt [1962].

*33. Let K be a finite distributive lattice and let L be the lattice of Theorem 17 as constructed in Lemmas 18 and 19. Show that L is sectionally complemented.
34. Let L be given as in exercise 33. Show that the congruences of L permute.
35. Let n be the length of the longest chain in K. Show that the L of Theorem 17 can be constructed so that the length of the longest chain is $2n - 1$ (define $a(b)$ only for $a >\!\!- b$).
36. Let P be a poset. Generalize Theorem 17 for $K = H_F(P)$ (notation is the same as in exercise 7.17).

Exercises 37 and 38 are from G. Grätzer and E. T. Schmidt [1958c].

*37. Show that a chain C is the congruence lattice of a lattice iff C is algebraic.
38. Prove that a Boolean algebra B is the congruence lattice of a lattice iff B is algebraic.
39. Let L be a distributive lattice. Show that $a \to \Theta[(a]]$ embeds L into $C(L)$.
40. Let L be a bounded distributive lattice. Show that for $a,b \in L$, $a \leq b$, $\Theta(a, b)$ has a complement in $C(L)$, namely, $\Theta(0, a) \vee \Theta(b, 1)$.

41. Let L be a bounded distributive lattice. Show that the compact elements of $C(L)$ form a Boolean lattice (J. Hashimoto [1952], G. Grätzer and E. T. Schmidt [1958f]).

42. Let B be a Boolean algebra generated by X and let L be the sublattice of B generated by X. Show that B is freely generated by X iff L is freely generated by X in **D** (see exercise 8.12).

43. Let the Boolean algebra B be generated by X. Show that B is freely generated by X iff $\bigwedge X_0 \leq \bigvee X_1$ implies that $X_0 \cap X_1 \neq \varnothing$ for any finite nonvoid $X_0, X_1 \subseteq X$ (compare this with Theorems 8.3 and 8.4).

10. Boolean Algebras R-generated by Distributive Lattices

The investigations of this section are based on the following result:

THEOREM 1. *Every distributive lattice can be embedded in a Boolean lattice.*

PROOF. By Theorem 7.19, every distributive lattice L is isomorphic to a ring of subsets of some set X. Obviously, L can be embedded into $P(X)$. ●

DEFINITION 2. *Let L be a sublattice of the generalized Boolean lattice B. The L is said to R-generate B if L generates B as a ring, and if L has a 0 (or 1) the same element is the 0 (or 1) of B.*

The last two conditions are equivalent to the following:

If $\bigwedge L$ exists, then $\bigwedge L = \bigwedge B$, and if $\bigvee L$ exists, then $\bigvee L = \bigvee B$.

Our goal is to show the uniqueness of the generalized Boolean lattice R-generated by L. The first result is essentially due to H. M. MacNeille [1939]:

LEMMA 3. *Let B be R-generated by L. Then every $a \in B$ can be expressed in the form*

$$a_0 + a_1 + \cdots + a_{n-1}, \qquad a_0 \leq a_1 \leq \cdots \leq a_{n-1}, a_0, \ldots, a_{n-1} \in L.$$

EXAMPLE. Let B be the Boolean lattice shown in Figure 10.1 and let $L = \{0, a_0, a_1, a_2\}$. Then L R-generates B.

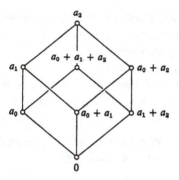

Figure 10.1

PROOF. Let B_1 denote the set of all elements that can be represented in the form $a_0 + \cdots + a_{n-1}$, $a_0, \ldots, a_{n-1} \in L$. Then $L \subseteq B_1$, and B_1 is closed under $+$ and $-$ (since $x - y = x + y$). Furthermore,

$$(a_0 + \cdots + a_{n-1})(b_0 + \cdots + b_{m-1}) = \sum a_i b_j,$$

and each term $a_i b_j = a_i \wedge b_j \in L$, so B_1 is closed under multiplication. We conclude that $B_1 = B$.

Note that L is a sublattice of B; therefore, for $a, b \in L$, $a \vee b$ in L is the same as $a \vee b$ in B. Thus $a \vee b = a + b + ab$, and so

$$a + b = ab + (a \vee b) = (a \wedge b) + (a \vee b).$$

Take $a_0 + \cdots + a_{n-1} \in B$. We prove by induction on n that the summands can be made to form an increasing sequence. For $n = 1$ this is obvious. Let us assume that $a_1 \leq \cdots \leq a_{n-1}$. Then

$a_0 + a_1 + \cdots + a_{n-1}$

$\quad = (a_0 \wedge a_1) + (a_0 \vee a_1) + a_2 + \cdots + a_{n-1}$

$\quad = (a_0 \wedge a_1) + ((a_0 \vee a_1) \wedge a_2) + (a_0 \vee a_2) + a_3 + \cdots + a_{n-1}$

$\quad = (a_0 \wedge a_1) + ((a_0 \vee a_1) \wedge a_2) + ((a_0 \vee a_2) \wedge a_3)$
$$\qquad\qquad + (a_0 \vee a_3) + \cdots + a_{n-1} = \cdots$$

$\quad = (a_0 \wedge a_1) + ((a_0 \vee a_1) \wedge a_2) + \cdots + ((a_0 \vee a_{n-2}) \wedge a_{n-1})$
$$\qquad\qquad + (a_0 \vee a_{n-1}),$$

and

$a_0 \wedge a_1 \leq (a_0 \vee a_1) \wedge a_2 \leq \cdots \leq (a_0 \vee a_{n-2}) \wedge a_{n-1} \leq a_0 \vee a_{n-1}.$ ●

LEMMA 4.   *Let L be a distributive lattice. Then there exists a generalized Boolean lattice B freely R-generated by L—that is, a generalized Boolean lattice B with the following properties:*

   (i) *B is R-generated by L;*
   (ii) *If B_1 is R-generated by L, then there is a homomorphism φ of B onto B_1 that is the identity map on L.*

PROOF.   The existence of B can be proved by copying the proof of Theorem 5.5 (Theorem 5.24), *mutatis mutandis.*   ●

LEMMA 5 (J. Hashimoto [1952]).   *Let B be a generalized Boolean lattice generated by L. Then every congruence relation of L has one and only one extension to B.*

PROOF.   The existence of an extension was proved in Theorem 9.6. By Theorems 9.7 and 9.10(i), the following statement implies the uniqueness of the extension:

If I and J are (ring) ideals of B with $I \subset J$, then there are elements $a,b \in L$, $a \neq b$, such that $a \equiv b \pmod{J}$ and $a \not\equiv b \pmod{I}$.

Indeed, let $x \in J - I$. By Lemma 3, x can be represented in the form

$$x = x_0 + \cdots + x_{n-1}, x_0 \leq \cdots \leq x_{n-1}, x_0, \cdots, x_{n-1} \in L.$$

If n is odd, then $x_0 = x \cdot x_0 \leq x \in J$, and thus $x_0 \in J$; $x_0 + x_1 + x_2 = x \cdot x_2 \in J$, therefore $x_1 + x_2 = x_0 + (x_0 + x_1 + x_2) \in J$. Similarly, $x_3 + x_4$, $x_5 + x_6, \ldots \in J$. Since $x_0 + (x_1 + x_2) + (x_3 + x_4) + \cdots \in J - I$, we conclude that either $x_0 \in J - I$, or $x_{2i} + x_{2i+1} \in J - I$ for some $2i < n$. If n is even, then we obtain $x_0 + x_1, x_2 + x_3, \ldots \in J$ (by multiplying x by $x_1, x_3, \ldots$), and we conclude that for some $2i < n$, $x_{2i} + x_{2i+1} \in J - I$.

Now if $x_{2i} + x_{2i+1} \in J - I$, then $x_{2i} \equiv x_{2i+1} \pmod{J}$, but $x_{2i} \not\equiv x_{2i+1} \pmod{I}$, $x_{2i}, x_{2i+1} \in L$. Finally, if $x_0 \in J - I$, then $x_0 \equiv 0 \pmod{J}$ and $x_0 \not\equiv 0 \pmod{I}$. This completes the proof, provided that $0 \in L$. If $0 \notin L$, then we can choose a $y \in I$ with $y < x_0$ and we obtain $y \equiv x_0 \pmod{J}$, $y \not\equiv x \pmod{I}$.   ●

THEOREM 6.   *If B_1 and B_2 are generalized Boolean lattices R-generated by a distributive lattice L, then B_1 and B_2 are isomorphic.*

REMARK.   For a distributive lattice L we shall denote by $B(L)$ a generalized Boolean lattice R-generated by L.

PROOF.   Let B be a free generalized Boolean lattice R-generated by L (see Lemma 4). Let φ be a homomorphism of B onto B_1 such that φ is the identity on L (see Lemma 4(ii)). We want to show that φ is an isomorphism. Indeed, if φ is not an isomorphism, then the ideal

$$I = \{x \mid x \in B, x\varphi = 0\}$$

is not 0. Thus by Lemma 5, $a \equiv b \pmod{I}$ for some $a,b \in L$, $a \neq b$. This means that $a\varphi = b\varphi$, contrary to our assumptions. Similarly, there is an isomorphism ψ between B and B_2. Obviously, $\varphi^{-1}\psi$ is an isomorphism between B_1 and B_2.   ●

COROLLARY 7.   *Let L_0 and L_1 be distributive lattices and let φ be a homomorphism of L_0 onto L_1 preserving 0 and/or 1, if they exist in L_0. Then φ can be extended to a homomorphism of $B(L_0)$ onto $B(L_1)$.*

PROOF.   Let Θ be the congruence relation of L_0 induced by φ ($x \equiv y(\Theta)$ iff $x\varphi = y\varphi$) and let $\overline{\Theta}$ be the extension of Θ to $B(L_0)$ (Lemma 5). Then $B(L_0)/\overline{\Theta}$ is a generalized Boolean lattice R-generated by $L_0/\Theta \cong L_1$. Thus $B(L_0)/\overline{\Theta} \cong B(L_1)$ by Theorem 6, and using this, the proof of Corollary 7 becomes trivial.   ●

COROLLARY 8.   *Let L_0 be a sublattice of the distributive lattice L_1 and assume that if L_1 has 0 and/or 1, then so does L_0, and the 0, 1 of L_1 is the 0, 1 of L_0. Let B denote the subalgebra of $B(L_1)$ R-generated by L_0. Then $B(L_0) \cong B$.*

PROOF.   The proof is trivial.   ●

Let L_0 and L_1 be given as in Corollary 8. It is natural to query the conditions under which L_0 R-generates $B(L_1)$. For $H \subseteq B(L_1)$, let $[H]_R$ denote the generalized Boolean sublattice of $B(L_1)$ R-generated by H. We can answer our query by determining $[L_0]_R \cap L_1$.

LEMMA 9.   *Let L_0 and L_1 be given as in Corollary 8 and let L_0 have a zero. Then $L_1 \cap [L_0]_R$ is the smallest sublattice of L_1 containing L_0 that is closed under taking relative complements in L_1. Therefore, L_0 R-generates*

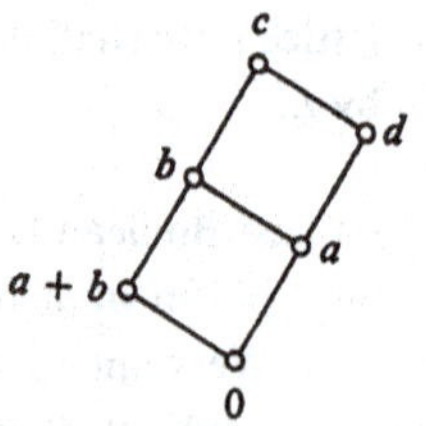

Figure 10.2

$B(L_1)$ iff the smallest sublattice of L_1 containing L_0 and closed under relative complementation in L_1 is L_1 itself.

PROOF.   It is obvious that $L_0 \subseteq L_1 \cap [L_0]_R$. If $a,b,c \in L_1 \cap [L_0]_R$, $d \in L_1$, and d is a relative complement of b in $[a, c]$, then $d = a + b + c \in L_1 \cap [L_0]_R$, since (see Figure 10.2) d is a relative complement of $a + b$ in the interval $[0, c]$. Thus $d \in L_1 \cap [L_0]_R$. Now suppose that L is a sublattice of L_1 containing L_0 and closed under relative complementation in L_1. If $x \in L_1 \cap [L_0]_R$, then by Lemma 3 we can represent x as

$$x = a_0 + \cdots + a_{n-1}, \qquad a_0, \ldots, a_{n-1} \in L_0, a_0 \leq \cdots \leq a_{n-1}.$$

We prove $x \in L$ by induction on n. If $n = 1$, $x = a_0 \in L_0 \subseteq L$. If $n = 2$, then x is the relative complement of a_0 in $[0, a_1]$, $0, a_0, a_1 \in L_0$, thus $x \in L$. If $n = 3$, then (see Figure 10.1) $x = a_0 + a_1 + a_2$ is the relative complement of a_1 in $[a_0, a_2]$, and so $x \in L$. Now let $n > 3$ and let $y \in L$ be proved for all $y = b_0 + \cdots + b_{k-1}$, $b_0, \ldots, b_{k-1} \in L_0$, $b_0 \leq \cdots \leq b_{k-1}$, and $k < n$. Note that $x \in L_1$ and $a_{n-3} \in L_0$ imply that $x a_{n-3} = a_0 + \cdots + a_{n-3} + a_{n-3} + a_{n-3} = a_0 + \cdots + a_{n-3} \in L_1$, and $x \vee a_{n-3} = x + a_{n-3} + x a_{n-3} = a_0 + \cdots + a_{n-1} + a_{n-3} + a_0 + \cdots + a_{n-3} = a_{n-3} + a_{n-2} + a_{n-1} \in L_1$. By the induction hypothesis, $a_0 + \cdots + a_{n-3} \in L$ and $a_{n-3} + a_{n-2} + a_{n-1} \in L$; therefore, x is the relative complement in L_1 of an element (namely, a_{n-3}) of L in an interval (namely, $[a_0 + \cdots + a_{n-3}, a_{n-3} + a_{n-2} + a_{n-1}]$) in L, and so, by assumption, $x \in L$. Thus $L_1 \cap [L_0]_R \subseteq L$. ●

In Theorem 1 we embedded L into $P(X)$, which is a complete Boolean lattice. The question arises whether we can require this embedding to be complete—that is, to preserve arbitrary meets and joins, if they exist in L.

It is easy to see that not every complete distributive lattice has a complete embedding into a complete Boolean lattice.

LEMMA 10 (J. von Neumann [1936]).   *Let B be a complete Boolean lattice. Then B satisfies the Join Infinite Distributive Identity* (*JID*),[4]

$$x \wedge \bigvee (x_i \mid i \in I) = \bigvee (x \wedge x_i \mid i \in I),$$

and its dual, the Meet Infinite Distributive Identity (*MID*).

PROOF. $x \wedge x_i \leq x$ and $x \wedge x_i \leq \bigvee (x_i \mid i \in I)$; therefore, $x \wedge \bigvee (x_i \mid i \in I)$ is an upper bound for $\{x \wedge x_i \mid i \in I\}$. Now let u be any upper bound, that is, $x \wedge x_i \leq u$ for all $i \in I$. Then

$$x_i = x_i \wedge (x \vee x') = (x_i \wedge x) \vee (x_i \wedge x') \leq u \vee x'.$$

Thus

$$x \wedge \bigvee (x_i \mid i \in I) \leq x \wedge (u \vee x') = (x \wedge u) \vee (x \wedge x') = x \wedge u \leq u,$$

showing that $x \wedge \bigvee (x_i \mid i \in I)$ is the least upper bound for $\{x \wedge x_i \mid i \in I\}$. (MID) follows by duality. ●

COROLLARY 11.   *Any complete distributive lattice that has a complete embedding into a complete Boolean lattice satisfies both* (*JID*) *and* (*MID*).

Easy examples show that (JID) and (MID) need not hold in all complete distributive lattices.

Our task now is to show the converse of Corollary 11 (N. Funayama [1959]). The construction depends on a property of $B(L)$ and on Theorem 6.4.

LEMMA 12   (V. Glivenko [1929]).   *Let L be a distributive lattice with 0. Then I(L) is a pseudocomplemented lattice in which*

$$I^* = \{x \mid x \wedge i = 0 \text{ for all } i \in I\}.$$

Let $S(I(L)) = \{I^* \mid I \in I(L)\}$. *If L is a Boolean lattice, then* $S(I(L))$ *is a complete Boolean lattice and the map* $a \to (a]$ *embeds L into* $S(I(L))$; *this embedding preserves all existing meets and joins.*

---

[4] Of course, (JID) is not an identity in the sense of Section 4 but is only an infinitary analogue of an identity.

PROOF. The first statement is trivial. Now let L be Boolean. It follows from Theorem 6.4 that $S(I(L))$ is a Boolean lattice. Furthermore, it is easily seen that for any $X \subseteq I(L)$, the inf and sup of X in $S(I(L))$ are $\bigwedge X$ and $(\bigvee X)^{**}$, respectively, where $\bigwedge$ and $\bigvee$ are the meet and join of X in $I(L)$, respectively. Since $\bigwedge ((x] \mid x \in X) = (\inf X]$, whenever $\inf X$ exists in L, the mapping $a \to (a]$ preserves all existing meets in L. Observe that, for $x, a \in L$, $x \wedge a' = 0$ iff $x \le a$, and so $(a] = (a']^* \in S(I(L))$. Now let $a = \sup X$ in L and set $I = (X] \,(= \bigvee ((x] \mid x \in X))$. To show that $x \to (x]$ is join-preserving, we have to verify that $I^{**} = (a]$, or, equivalently, that $I^* = (a']$. Indeed, if $b \in I^*$, then $b \wedge x = 0$ for all $x \in I$, and thus $x \le b'$. Therefore, $a = \sup X \le b'$, proving $a' \ge b$, that is, $b \in (a']$. Conversely, let $b \in (a']$. Then $b' \ge a$; therefore, $b' \ge a = \sup X \ge x$ for all $x \in X$, and so $b \wedge x = 0$ for all $x \in X$. This shows that $b \in I^*$, proving that $I^* = (a']$. ●

LEMMA 13. *Let L be a complete lattice satisfying (JID) and (MID). Then the identity map is a complete embedding of L into $B(L)$.*

PROOF. Let us write $a \in B(L)$ in the form

$$a = a_0 + \cdots + a_{n-1}, \qquad a_0 \le a_1 \le \cdots \le a_{n-1}, a_0, \ldots, a_{n-1} \in L.$$

Since L is complete, it has a 0, and thus (writing a_0 as $0 + a_0$) we can assume without loss of generality that n is odd. We claim that for $x \in L$ and $a \in B(L)$, we have $x \le a$ iff

$$x \wedge a_0 = x \wedge a_1,$$

$$x \wedge a_2 = x \wedge a_3, \ldots,$$

$$x \wedge a_{n-3} = x \wedge a_{n-2}, \qquad x \le a_{n-1}.$$

Indeed, let $x \le a$. Then

$$xa_1 = xa_1(a_0 + \cdots + a_{n-1}) = x(a_0 + a_1 + a_1 + \cdots + a_1) = xa_0;$$

therefore, $x \wedge a_0 = x \wedge a_1$. Thus $x(a_2 + \cdots + a_{n-1}) = (xa_0 + xa_1) + x(a_2 + \cdots + a_{n-1}) = xa = x$, and so $x \le a_2 + \cdots + a_{n-1}$. Conversely, if $x \wedge a_0 = x \wedge a_1$ and $x \le a_2 + \cdots + a_{n-1}$, then $xa = xa_0 + xa_1 + x(a_2 + \cdots + a_{n-1}) = x$, proving that $x \le a$. A simple induction completes the proof of the claim.

Let $X \subseteq L$, $y = \sup X$ in L, and $a \in B(L)$. If $x \le a$ for all $x \in X$, then the formulas of the preceding claim hold for all x and a and, by (JID), for y and a, proving that $y \le a$. Thus $y = \sup X$ in $B(L)$. The dual argument, using (MID), completes the proof. ●

THEOREM 14 (N. Funayama [1959]).   *A complete lattice L has a complete embedding into a complete Boolean lattice iff L satisfies (MID) and (JID).*

PROOF.   Combine Lemmas 10–13. ●

The representation for $a \in B(L)$ given in Lemma 3 is not unique in general; the only exception is when L is a chain. Since this case is of special interest, we shall investigate it in detail.

Repeating the definition, a Boolean lattice B is *R-generated by a chain* if $B = B(C)$ for some chain $C \subseteq B$. This concept is due to M. Mostowski and A. Tarski [1939] and can be extended to distributive lattices as follows:

A distributive lattice L is *R-generated by a chain* $C (\subseteq L)$ if C R-generates $B(L)$.

The following notation will facilitate the proof of the next lemma as well as the discussion at the end of this section. For a chain C in a lattice L, write C^0 to denote the chain C if L has no zero and let C^0 denote $C \cup \{0\}$ if L has a zero.

LEMMA 15.   *Let L be a distributive lattice and let C be a chain in L. Then C R-generates L iff L is the smallest sublattice of itself containing C^0 and closed under the formation of relative complements.*

PROOF.   Apply Lemma 9 to C^0. ●

An explicit representation of $B(C)$ is given as follows:

For a chain C, let $B[C]$ be the set of all subsets of C of the form

$$(a_0] + (a_1] + \cdots + (a_{n-1}], \quad 0 < a_0 \le a_1 \le \cdots \le a_{n-1},$$

$$a_0, \ldots, a_{n-1} \in C,$$

and $\varnothing$, where $+$ is the symmetric difference. We consider $B[C]$ as a poset, partially ordered by $\subseteq$. The clause "$0 < a_0$" means that $0 < a_0$ if 0 exists; otherwise there is no restriction on a_0.

We identify $a \in C$ with $(a]$, for $a \ne 0$, and 0 (if it exists) with $\varnothing$. Thus $C \subseteq B[C]$.

LEMMA 16. $B[C]$ *is the generalized Boolean lattice R-generated by* C.

PROOF.   The proof is obvious, by construction and by Theorem 6.   ●
Note that every nonvoid element a can be represented in the form

$$a = (a_0] \cup (b_1, a_1] \cup \cdots \cup (b_{n-1}, a_{n-1}],$$

$$0 < a_0 \le b_1 \le a_1 \le \cdots \le b_{n-1} \le a_{n-1},$$

where the union is disjoint union, the first term $(a_0]$ may be missing, and $(x, y]$ stands for $\{t \mid x < t \le y\}$. An element of the form $(x, y]$ is nothing but $x + y$. Thus, $a = a_0 + b_1 + a_1 + \cdots + b_{n-1} + a_{n-1}$, and so we conclude:

COROLLARY 17.   *In* $B(C)$ *every nonzero element* a *has a unique representation in the form*

$$a = a_0 + a_1 + \cdots + a_{n-1},$$

$$0 < a_0 < a_1 < \cdots < a_{n-1}, a_0, \ldots, a_{n-1} \in C.$$

The following results show that many distributive lattices can be R-generated by chains.

LEMMA 18.   *Every finite Boolean lattice* B *can be R-generated by a chain; in fact,* $B = [C]_R$ *for any maximal chain* C *of* B.

PROOF.   Let B_1 be the subalgebra of B R-generated by C. Using the notation of Corollary 7.14, the length of $C = |J(B)|$; also, the length of $C = |J(B_1)|$; thus $|J(B)| = |J(B_1)| = n$. We conclude that both B and B_1 have 2^n elements, proving that $B = B_1$.   ●

COROLLARY 19.   *Every finite distributive lattice* L *can be R-generated by a chain—in fact, by any maximal chain of* L.

PROOF.   Let C be a maximal chain in L and let $B = B(L)$. Then $|J(L)| = |J(B)|$. By Corollary 7.14, C is maximal in B. Thus, $B = B(C) \supseteq L$.   ●

THEOREM 20.   *Every countable distributive lattice* L *can be R-generated by a chain.*

PROOF. Let $L = \{a_0, a_1, a_2, \ldots, a_n, \ldots\}$ and let L_n be the sublattice of L generated by $a_0, \ldots, a_n$. Let C_0 be a maximal chain of L_0, and, inductively, let C_n be a maximal chain of L_n containing C_{n-1}. Set $C = \bigcup (C_i \mid i < \omega)$. We claim that C generates L. Take $a \in B(L)$; $a = x_0 + \cdots + x_{m-1}$, $x_0, \ldots, x_{m-1} \in L$. $L = \bigcup (L_i \mid i < \omega)$; thus for some n, $x_0, \ldots, x_{m-1} \in L_n$, and so $a \in B(L_n)$. Since L_n is finite, we get $B(L_n) = B(C_n)$; therefore $a \in B(C_n) \subseteq B(C)$, proving that $L \subseteq B(C)$. ●

COROLLARY 21.   *The correspondence $C \to B(C)$ maps the class of countable chains onto the class of countable generalized Boolean lattices. Under this correspondence, subchains and homomorphic images correspond to subalgebras and homomorphic images.*

Note, however, that $C_0 \cong C_1$ is *not* implied by $B(C_0) \cong B(C_1)$.

Much is known about countable chains. Utilizing the previous results, such information can be used to prove results on countable generalized Boolean lattices. To help distinguish an important class of chains, we introduce the concept of *prime interval*. An interval $[a, b]$ is prime if $a \prec b$.

LEMMA 22.   *Every countable chain C can be embedded in the chain Q of rational numbers. Every countable chain not containing any prime interval is isomorphic to one of the intervals $(0, 1)$, $[0, 1)$, $(0, 1]$, and $[0, 1]$ of Q.*

PROOF. Let $C = \{x_0, x_1, \ldots, x_{n-1} \cdots\}$. We define the map φ inductively as follows: Pick an arbitrary $r_0 \in Q$ and set $x_0\varphi = r_0$. If $x_0\varphi, \ldots, x_{n-1}\varphi$ have already been defined, we define $x_n\varphi$ as follows: Let

$$L_n = \bigcup ((x_i\varphi] \mid x_i < x_n, i < n)$$
$$U_n = \bigcup ([x_i\varphi) \mid x_i > x_n, i < n)$$

($L_n = \varnothing$ or $U_n = \varnothing$ is possible). Note that if $L_n \neq \varnothing$, then it has a greatest element l_n, and if $U_n \neq \varnothing$, then it has a smallest element u_n. If both are nonempty, then $l_n < u_n$. In any case, we can choose an $r_n \in Q$ with $r_n \notin L_n \cup U_n$. We set $x_n\varphi = r_n$. Obviously, φ is an embedding.

The second part of the proof reduces to the following statement:

Let C and D be bounded countable chains with no prime intervals. Then $C \cong D$.

To prove this, let $C = \{c_0, c_1, \ldots\}$ and $D = \{d_0, d_1, \ldots\}$. We define two maps: $\varphi: C \to D$, $\psi: D \to C$. Let us assume that $c_0 = 0$, $c_1 = 1$, and that

$d_0 = 0$, $d_1 = 1$. For each $n < \omega$, we shall define inductively finite chains C_n, D_n ($C_n \subseteq C$, $D_n \subseteq D$) and an isomorphism $\varphi_n \colon C_n \to D_n$ with inverse $\psi_n \colon D_n \to C_n$. Set $C_0 = \{c_0, c_1\} = \{0, 1\}$, $D_0 = \{d_0, d_1\} = \{0, 1\}$ and $i\varphi_0 = i$, $i\psi_0 = i$, for $i = 0, 1$. Given C_n, D_n, φ_n, ψ_n, and n even, let k be the smallest integer with $c_k \notin C_n$. Define $u_k = \bigwedge ([c_k) \cap C_n)$ and $l_k = \bigvee ((c_k] \cap C_n)$. Then $l_k < c_k < u_k$, and so $l_k \varphi_n < u_k \varphi_n$. Since D contains no prime intervals, we can choose a $d \in D$ satisfying $l_k \varphi_n < d < u_k \varphi_n$. Since ψ_n is isotone, $d \notin D_n$. Define $C_{n+1} = C_n \cup \{c_k\}$, $D_{n+1} = D_n \cup \{d\}$, φ_{n+1} restricted to C_n to be φ_n, and $c_k \varphi_{n+1} = d$, ψ_{n+1} restricted to D_n to be ψ_n and $d\psi_{n+1} = c_k$. If n is odd we proceed in a similar way, but we interchange the role of C and D, C_n and D_n, φ_n and ψ_n, respectively.

Finally, put $\varphi = \bigcup (\varphi_n \mid n < \omega)$. Clearly, $C = \bigcup (C_n \mid n < \omega)$, $D = \bigcup (D_n \mid n < \omega)$, and φ is the required isomorphism. ●

COROLLARY 23.   *Up to isomorphism there is exactly one countable Boolean lattice with no atoms and exactly one countable generalized Boolean lattice with no atoms and no unit element, $B(D)$, where D is the $[0, 1)$ rational interval.*

PROOF.   Take the rational intervals $[0, 1]$ and $[0, 1)$. The generalized Boolean lattices in question are $B([0, 1])$ and $B([0, 1))$. This follows from the observation that $[a, b]$ is a prime interval in C iff $a + b$ is an atom in $B(C)$. The results follow from Lemmas 16 and 22 and Theorem 20. ●

THEOREM 24.   *Let B be a countable Boolean algebra. Then B has either $\aleph_0$ or $2^{\aleph_0}$ prime ideals.*

REMARK.   This is obvious if we assume the Continuum Hypothesis. Interestingly enough, we can give a proof without it.

PROOF.   For a Boolean algebra B and an ideal I of B, we shall write B/I for $B/\Theta[I]$. If J is an ideal of B with $J \supseteq I$, then $J/I = \{[x]\Theta[I] \mid x \in J\}$ is an ideal of B/I. (This is the usual notation in ring theory.) Let B be a Boolean algebra. We define the ideals I_γ by transfinite induction. $I_0 = (0]$; I_1 is the ideal generated by the atoms of B; given I_γ, let I be the ideal of B/I_γ generated by the atoms of B/I_γ and let $\varphi \colon x \to x + I_\gamma$ be the homomorphism of B onto B/I_γ; we set $I_{\gamma+1} = I\varphi^{-1}$. Finally, if γ is a limit ordinal, set $I_\gamma = \bigcup (I_\delta \mid \varphi < \gamma)$. The *rank* of B is defined to be the smallest ordinal α such that $I_\alpha = I_{\alpha+1}$. Obviously, $\bar\alpha \leq |B|$.

CLAIM 1.   *Let B be countable. If $I_\alpha \neq B$, then $|\mathscr{P}(B)| = 2^{\aleph_0}$.*

Indeed, if $I_\alpha \neq B$, then B/I_α has no atoms, and thus $B/I_\alpha \cong B(C)$, where C is the rational interval $[0, 1]$. By Lemma 5 (see exercise 27), $|\mathscr{P}(B/I_\alpha)| = |\mathscr{P}(C)| = |I(C)| = 2^{\aleph_0}$.

CLAIM 2.   *Let B be countable. If $I_\alpha = B$, then $|\mathscr{P}(B)| = \aleph_0$.*

Indeed, for $\gamma < \alpha$, let $\mathscr{P}_\gamma(B)$ be the set of prime ideals P of B for which $I_\gamma \subseteq P$, $I_{\gamma+1} \nsubseteq P$. Since $\bar{a} \leq \aleph_0$, it suffices to show that $|\mathscr{P}_\gamma(B)| = \aleph_0$. If $P \in \mathscr{P}_\gamma(B)$, then, by Corollary 7.16 and Theorem 7.22, we have $P \vee I_{\gamma+1} = B$. It follows that for $P, Q \in \mathscr{P}_\gamma(B)$, $P \neq Q$, we have $P \cap I_{\gamma+1} \neq Q \cap I_{\gamma+1}$. Thus $P \to (P \cap [I_{\gamma+1}]_R)/I_\gamma$ is a one-to-one correspondence of $\mathscr{P}_\gamma(B)$ into (in fact, onto) $\mathscr{P}([I_{\gamma+1}]_R/I_\gamma)$; but $[I_{\gamma+1}]_R/I_\gamma$ is just the generalized Boolean lattice of all finite subsets of a countable set. Therefore, $|\mathscr{P}_\gamma(B)| = \aleph_0$. ●

In order to avoid giving the impression that most Boolean algebras can be R-generated by chains, we state

LEMMA 25.   *Let B be a complete Boolean algebra R-generated by a chain C. Then B is finite.*

PROOF.   Let $B = [C]_R$ and let the chain C be infinite. We can assume that C contains a subchain $\{x_0, x_1, \ldots, x_n, \ldots\}$, $x_0 < x_1 < \cdots < x_n < \cdots$ (or dually, in which case we dualize the proof). Then define $a_n = x_0 + x_1 + \cdots + x_{2n} + x_{2n+1}$ for each $n < \omega$. We claim that $\bigvee (a_i \mid i < \omega)$ does not exist. Indeed, let a be an upper bound for $\{a_i \mid i < \omega\}$. By the remarks immediately following Lemma 16, we can represent each a_n by a set $(x_0, x_1] \cup (x_2, x_3] \cup \cdots \cup (x_{2n}, x_{2n+1}]$, and we can represent a in the form

$$a = (b_0] \cup (b_1, b_2] \cup \cdots \cup (b_{m-2}, b_{m-1}],$$

where $0 < b_0 < b_1 < \cdots < b_{m-1}$, $b_i \in C$ for $i < m$, and the first term, $(b_0]$, may be missing. Since a contains each a_n, there must exist an n and a $j < m$ such that both $(x_{2n}, x_{2n+1}]$ and $(x_{2n+2}, x_{2n+3}]$ are contained in $(b_{j-1}, b_j]$ (or in $(b_0]$ if $j = 0$). Therefore, the interval $(x_{2n+1}, x_{2n+2}]$ can be deleted from a, and it will still contain all the a_i—that is, $a + x_{2n+1} + x_{2n+2}$ is an upper bound for $\{a_i \mid i < \omega\}$, and $a + x_{2n+1} + x_{2n+2} < a$. We conclude that $\{a_i \mid i < \omega\}$ does not have a least upper bound. ●

Next we consider which chains can be R-generating chains of a given distributive lattice.

DEFINITION 26. *Let L be a distributive lattice and let C be a chain in L. The chain C is called* strongly maximal *in L iff, for any homomorphism φ of L onto a distributive lattice L_1, the chain $(C\varphi)^0$ is maximal in L_1. (The notation C^0 was introduced preceding Lemma 15.)*

LEMMA 27. *If the distributive lattice L is R-generated by a chain $C \subseteq L$, then C^0 is maximal in L.*

PROOF. If C^0 is not maximal in L, then we can find $a \in L$, $a \notin C$, $a \neq 0$, such that $C \cup \{a\}$ is a chain. Write $a = a_0 + a_1 + \cdots + a_{n-1}$ with $0 < a_0 < a_1 < \cdots < a_{n-1}$ and $a_i \in C$ for $i < n$. Since $a \notin C$, $n > 1$. Now $a \wedge a_0 = a_0 + a_0 + \cdots + a_0$, which is a_0 if n is odd and 0 if n is even. But $a_0 \neq a$ and $0 \neq a$, so, since a and a_0 are comparable, $a \wedge a_0 = a_0$, and n is odd. Then $a \wedge a_1 = a_0 + a_1 + \cdots + a_1 = a_0$, contradicting the comparability of a and a_1. ●

The converse of Lemma 27 is false. The following theorem settles the matter.

THEOREM 28. *Let L be a distributive lattice and let C be a chain in L. Then C R-generates L iff C is strongly maximal in L.*

PROOF. If C R-generates L, then, for any homomorphism φ, $C\varphi$ R-generates $L\varphi$. By Lemma 27, $(C\varphi)^0$ is maximal in $L\varphi$, so C is strongly maximal in L.

Next assume that C is strongly maximal in L but does not R-generate L. Without any loss of generality we can assume that L and C have a greatest element. (Otherwise, add one. Then $C \cup \{1\}$ is strongly maximal in $L \cup \{1\}$ but does not R-generate $L \cup \{1\}$.) Let $B_1 = B(L)$ and let $B_0 = [C]_R$. By hypothesis, $B_0 \neq B_1$, so there exists an $a \in B_1 - B_0$. We claim that there exist prime ideals $P_1 \neq P_2$ of B_1 with $B_0 \cap P_1 = B_0 \cap P_2$. With $I = ((a] \cap B_0]$ and $D = [a)$, we have $I \cap D = \varnothing$, so, by Theorem 7.15, there is a prime ideal P_1 such that $I \subseteq P_1$ and $P_1 \cap D = \varnothing$. Then let $I_1 = (a]$ and $D_1 = [B_0 - P_1)$. Since $(a] \cap B_0 \subseteq P_1$, it follows that $I_1 \cap D_1 = \varnothing$. Let P_2 be a prime ideal with $I_1 \subseteq P_2$, $P_2 \cap D_1 = \varnothing$. Then $a \in P_2 - P_1$, so $P_1 \neq P_2$. Because $P_2 \cap (B_0 - P_1) = \varnothing$, $P_2 \cap B_0 \subseteq P_1 \cap B_0$. Then by

Theorem 7.22 (prime ideals of a Boolean lattice are unordered), $P_1 \cap B_0 = P_2 \cap B_0$, proving our claim.

Now (again by using Theorem 7.22) we can map B_1 onto $(\mathfrak{C}_2)^2$ by a homomorphism ψ: For $x \in P_1 \cap P_2$, $x\psi = \langle 0, 0 \rangle$; for $x \in P_1 - P_2$, $x\psi = \langle 0, 1 \rangle$; for $x \in P_2 - P_1$, $x\psi = \langle 1, 0 \rangle$; for $x \notin P_1 \cup P_2$, $x\psi = \langle 1, 1 \rangle$. Since $C\psi = \{ \langle 0, 0 \rangle, \langle 1, 1 \rangle \}$ is not maximal and $C\psi = (C\psi)^0$, we conclude that C is not strongly maximal in L. ●

COROLLARY 29.   *Let C and D be strongly maximal chains of the distributive lattice L. Then $|C| = |D|$ and $|I(C)| = |I(D)|$.*

PROOF.   If L is finite, these conclusions follow from Corollary 7.14. If L is infinite, then $[C] = [D] = B(L)$, and so $|C| = |D| = |L|$. Also, by Lemma 5, $|\mathcal{P}(C)| = |\mathcal{P}(B(L))| = |\mathcal{P}(D)|$, and $\mathcal{P}(C) = I(C)$, $\mathcal{P}(D) = I(D)$; hence the second statement. ●

Corollary 29 is the strongest known extension of Corollary 7.14 to the infinite case. The second part of Corollary 29 is from G. Grätzer and E. T. Schmidt [1957a].

Boolean algebras generated by chains were first investigated by M. Mostowski and A. Tarski [1939]. Theorem 20 for Boolean lattices and Theorem 24 were communicated to the author by J. R. Büchi. These results have been known for some time in topology (via the Stone topological representation theorem, see Section 11). Some of the other results are apparently new.

Exercises

1. Give a detailed proof of Lemma 4.
2. Try to describe the most general situation to which the idea of the proof of Theorem 5.5 (Theorem 5.24) could be applied.
3. Show that Lemma 5 does not remain valid if the word "generalized" is omitted.
4. Find necessary and sufficient conditions on L in order that L have a Boolean extension B to which every congruence of L has exactly one extension.
5. Let L be a distributive lattice and define L_1 to be the lattice L if L has a unit element; let L_1 be L with a unit added if L does not have a unit element. The *Boolean lattice $B[L]$ R-generated by L* is defined to be

$B(L_1)$. Show that if B is any Boolean lattice, containing L as a sublattice, and B is generated by L under $\wedge$, $\vee$, and $'$, then B is isomorphic to the Boolean lattice R-generated by L.

6. Work out Corollaries 7 and 8 for the Boolean lattice R-generated by L.

*7. The *Complete Infinite Distributive Identity* is (for $I, J \neq \varnothing$):

$$\bigwedge (\bigvee (a_{ij} \mid j \in J) \mid i \in I) = \bigvee (\bigwedge (a_{i\,i\varphi} \mid i \in I) \mid \varphi: I \to J).$$

Show that this holds in a complete Boolean lattice B iff it is atomic (A. Tarski [1930]). (Hint: apply the identity to

$$\bigwedge (a \vee a' \mid a \in B) = 1).$$

8. Prove that the Complete Infinite Distributive Identity is self-dual for Boolean lattices.

9. For a subset A of a lattice L, set

$$A^u = \{x \mid x \in L, x \text{ is an upper bound of } A\}$$
$$A^l = \{x \mid x \in L, x \text{ is a lower bound of } A\}.$$

Prove that $(A^u)^l \supseteq A$, $(A^l)^u \subseteq A$, $A^u = ((A^u)^l)^u$, and $A^l = ((A^l)^u)^l$.

10. Call an ideal I of a lattice L *normal* if $I = (I^u)^l$. Show that every principal ideal is normal.

11. Let $I_N(L)$ denote the set of all normal ideals of L. Show that $I_N(L)$ is a complete lattice but that it is not necessarily a sublattice of $I_0(L)$.

12. Show that the map: $x \to (x]$ is an embedding of L into $I_N(L)$, preserving all meets and joins that exist in L. ($I_N(L)$ is called the *MacNeille completion* of L; see H. M. MacNeille [1937].)

13. Let B be a Boolean lattice and let I be an ideal of B. Show that I is normal iff $I = I^{**}$ (for the concepts, see exercise 10 and Lemma 12).

14. Prove that the Boolean lattice $S(I(L))$ of Lemma 12 is the MacNeille completion of the Boolean lattice L.

*15. Show that the MacNeille completion of a distributive lattice need not even be modular.

16. Let L be a distributive algebraic lattice. Show that L satisfies the Join Infinite Distributive Identity. (Thus, for any lattice K, $C(K)$ satisfies (JID).)

17. Let L be a distributive lattice, $a_i, b_i \in L$ for $i < \omega$ and $[a_0, b_0] \supset [a_1, b_1] \supset \cdots$. Define

$$\Theta = \bigvee (\Theta(a_0, a_i) \vee \Theta(b_0, b_i) \mid i < \omega).$$

Show that

$$\Theta \vee \bigwedge (\Theta(a_i, b_i) \mid i < \omega) \neq \bigwedge (\Theta \vee \Theta(a_i, b_i) \mid i < \omega).$$

18. Use exercise 17 to show that, for a distributive lattice L, the Meet Infinite Distributive Identity holds in $C(L)$ iff every interval in L is finite (G. Grätzer and E. T. Schmidt [1958c]).

19. For a chain C, introduce and describe $B(C)$ using Corollary 17.

*20. Prove the converse of Lemma 18: If every maximal chain generates the Boolean lattice B, then B is finite.

21. Why is it not possible to use transfinite induction to extend Theorem 20 to the uncountable case?

22. Let C be a chain with 0 and 1 and let $a \in C$. Define a new order on C: For $x,y \le a$, and $a \le x,y$, let $x \le y$ retain its meaning; for $x \le a \le y$, let $y \le x$; let C_1 be the set C with the new order. Then C_1 is a chain, and $B(C) \cong B(C_1)$, but in general $C \cong C_1$ does not hold.

23. Describe a countable family of pairwise nonisomorphic countable Boolean algebras.

24. Let C be a maximal chain of the distributive lattice L. Prove that C is a maximal chain in $B(L)$.

25. Let L_0 be the $[0, 1]$ rational interval and let L_1 be the $[0, 1]$ real interval. Let $C = \{\langle x, x \rangle \mid 0 \le x \le 1, x \text{ rational}\}$. Then C is a maximal chain in $L_0 \times L_1$. Show that C is not strongly maximal (G. Grätzer and E. T. Schmidt [1957a]).

26. Find in $L_0 \times L_1$ of exercise 25 maximal chains of cardinality $\aleph_0$ and $2^{\aleph_0}$. What are the cardinalities of strongly maximal chains?

27. Let $B = B(L)$. Show that $P \to P \cap L$ for $P \in \mathscr{P}(B)$ is a one-to-one correspondence between the prime ideals of L and B (use Lemma 5).

*28. Let A be an infinite set, $B = P(A)$. Prove that B has maximal chains of cardinality $|A|$ and $2^{|A|}$.

29. Construct an example in which the sequence of ideals I_γ of Theorem 24 does not terminate in finitely many steps.

30. Let C be the $[0, 1]$ interval of the rational numbers. Show that $B(C)$ is $F_B(\aleph_0)$.

11. Topological Representation

The poset $\mathscr{P}(L)$ of prime ideals does give a great deal of information about the distributive lattice L, but obviously it does not characterize L. For instance, for a countably infinite Boolean algebra L, $\mathscr{P}(L)$ is an unordered set of cardinality $\aleph_0$ or $2^{\aleph_0}$, whereas there are surely more than two such Boolean algebras up to isomorphism.

Therefore, it is necessary to endow $\mathscr{P}(L)$ with more structure if we want it to characterize L. M. H. Stone [1937] endowed $\mathscr{P}(L)$ with a topology (see also L. Rieger [1949]). In this section we shall discuss his approach in a slightly more general but, in our opinion, more natural framework.

A join-semilattice L is called *distributive*[5] if $a \le b_0 \vee b_1 \ (a,b_0,b_1 \in L)$

---

[5] This concept appeared quite naturally in the research of G. Grätzer and E. T. Schmidt on congruence lattices of lattices in 1960–1961; because the research was not very successful, this concept did not appear in print until the late sixties.

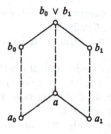

Figure 11.1

implies the existence of $a_0, a_1 \in L$, $a_i \leq b_i$, $i = 0,1$, with $a = a_0 \vee a_1$ (see Figure 11.1). Note that a_0 and a_1 need not be unique.

Some elementary properties of a distributive semilattice are as follows:

LEMMA 1.

(i) *If* $\langle L; \wedge, \vee \rangle$ *is a lattice, then the join-semilattice* $\langle L; \vee \rangle$ *is distributive iff the lattice* $\langle L; \wedge, \vee \rangle$ *is distributive.*

(ii) *If a join-semilattice* L *is distributive, then for any* $a,b \in L$ *there is a* $d \in L$ *with* $d \leq a$, $d \leq b$. *Consequently,* $I(L)$ *is a lattice.*

(iii) *A join-semilattice* L *is distributive iff* $I(L)$, *as a lattice, is distributive.*

PROOF.

(i) If $\langle L; \wedge, \vee \rangle$ is distributive, and $a \leq b_0 \vee b_1$, then with $a_i = a \wedge b_i$, $i = 0,1$, $a = a_0 \vee a_1$. Conversely, if $\langle L; \vee \rangle$ is distributive, and the lattice L has $\mathfrak{M}_5$ or $\mathfrak{N}_5$ as a sublattice, then $a \leq b \vee c$ (see notation of Figure 7.1) leads to a contradiction. ▶

(ii) $a \leq a \vee b$, thus $a = a_0 \vee b_0$, where $a_0 \leq a$, $b_0 \leq b$. Since, in addition, $b_0 \leq a$, b_0 is a lower bound for a and b. ▶

(iii) First we observe that, for $I, J \in I(L)$,

$$I \vee J = \{i \vee j \mid i \in I, j \in J\}$$

follows from the assumption that the join-semilattice L is distributive. Therefore, the distributivity of $I(L)$ can be easily proved. Conversely, if $I(L)$ is distributive, and $a \leq b_0 \vee b_1$, then

$$(a] = (a] \wedge ((b_0] \vee (b_1]) = ((a] \wedge (b_0]) \vee ((a] \wedge (b_1]),$$

and so $a = a_0 \vee a_1$, $a_0 \in (b_0]$, $a_1 \in (b_1]$, which is distributivity for L. ●

A subset D of a join-semilattice L is called a *dual ideal* if $a \in D$ and $x \geq a$ imply that $x \in D$, and $a,b \in D$ implies that there exists a lower bound d of $\{a, b\}$ such that $d \in D$. An ideal I of L is *prime* if $I \neq L$ and $L - I$ is a dual ideal. Again, let $\mathscr{P}(L)$ denote the set of all prime ideals of L.

LEMMA 2.   *Let I be an ideal and let D be a dual ideal of a distributive join-semilattice L. If $I \cap D = \varnothing$, then there exists a prime ideal P of L with $P \supseteq I, P \cap D = \varnothing$.*

PROOF.   The proof is a routine modification of the proof of Theorem 7.15.   ●

In the rest of this section, let L stand for a distributive join-semilattice with zero.

In $\mathscr{P}(L)$, sets of the form $r(a) = \{P \mid a \notin P\}$ represent the elements of L. We will make all these sets open.

Let $\mathscr{S}(L)$ denote the topological space defined on $\mathscr{P}(L)$ by postulating that the sets of the form $r(a)$ be a subbase [6] for the open sets; we shall call $\mathscr{S}(L)$ the *Stone space* of L.

LEMMA 3.   *Let I be an ideal of L,*

$$r(I) = \{P \mid P \in \mathscr{S}(L), P \nsupseteq I\}.$$

Then $r(I)$ is open in $\mathscr{S}(L)$. Conversely, every open set U of $\mathscr{S}(L)$ can be uniquely represented as $r(I)$ for some ideal I of L.

PROOF.   We simply observe that $r(I) \cap r(J) = r(I \wedge J)$,

$$r(\bigvee (I_j \mid j \in K)) = \bigcup (r(I_j) \mid j \in K),$$

and $r((a]) = r(a)$, from which it follows that the $r(I)$ form the smallest collection of sets closed under finite intersection and arbitrary union containing all the $r(a)$. Observe that $a \in I$ iff $r(a) \subseteq r(I)$. Thus $r(I) = r(J)$ iff $a \in I$ is equivalent to $a \in J$, that is, iff $I = J$.   ●

LEMMA 4.   *The subsets of $\mathscr{S}(L)$ of the form $r(a)$ can be characterized as compact open sets.*

---

[6] Exercises 1–22 review all the topological concepts that are used in this and the next section.

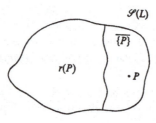

Figure 11.2

PROOF. Indeed, if a family of open sets $\{r(I_k) \mid k \in K\}$ is a cover for $r(a)$, that is, $r(a) \subseteq \bigcup (r(I_k) \mid k \in K) = r(\bigvee (I_k \mid k \in K))$, then $a \in \bigvee(I_k \mid k \in K)$. This implies that $a \in \bigvee (I_k \mid k \in K_0)$ for some finite $K_0 \subseteq K$, proving that $r(a) \subseteq \bigcup (r(I_k) \mid k \in K_0)$. Thus $r(a)$ is compact. Conversely, if I is not principal, then $r(I) \subseteq \bigcup (r(a) \mid a \in I)$, but $r(I) \nsubseteq \bigcup (r(a) \mid a \in I_0)$ for any finite $I_0 \subseteq I$. ●

From Lemma 4 we immediately conclude:

THEOREM 5. *The Stone space $\mathscr{S}(L)$ determines L up to isomorphism.*

If we want to use Stone spaces to construct new ones in order that new distributive lattices can be constructed from given ones, then we have to know what Stone spaces look like. Stone spaces are characterized in Theorem 8. To prepare for the proof of Theorem 8, we prove Lemma 6.

Let P be a prime ideal of L. Then P is represented as an element of $\mathscr{S}(L)$ and also by $r(P)$. The connection between P and $r(P)$ is given in Lemma 6 and is illustrated by Figure 11.2.

LEMMA 6. *For every prime ideal P of L, $\overline{\{P\}} = \mathscr{S}(L) - r(P)$, where $\overline{\{P\}}$ is the topological closure of $\{P\}$.*

PROOF. By definition of closure,

$$\overline{\{P\}} = \{Q \mid Q \in r(a) \quad \text{implies that} \quad P \in r(a)\} = \{Q \mid Q \supseteq P\}$$
$$= \mathscr{S}(L) - \{Q \mid Q \nsupseteq P\} = \mathscr{S}(L) - r(P). ●$$

COROLLARY 7. *If $P \neq Q$, then $\overline{\{P\}} \neq \overline{\{Q\}}$; in other words, $\mathscr{S}(L)$ is a T_0-space.*

PROOF. Combine Lemmas 3 and 6. ●

Lemma 6 also shows that if P is a prime ideal, then $\mathscr{S}(L) - r(P)$ must be the closure of a singleton. In other words:

(C)   If U is an open set with the property that, for the compact open sets U_0 and U_1, $U_0 \cap U_1 \subseteq U$ implies that $U_0 \subseteq U$ or $U_1 \subseteq U$, then $\mathscr{S}(L) - U = \overline{\{P\}}$ for some element P.

Now we can state the characterization theorem:

THEOREM 8.   *The Stone space $\mathscr{S}$ of a distributive join-semilattice with zero can be characterized (up to homeomorphism) by the following two properties:*

(S1)   *$\mathscr{S}$ is a T_0-space in which the compact open sets form a base for the open sets.*

(S2)   *If F is a closed set in $\mathscr{S}$, $\{U_k \mid k \in K\}$ is a dually directed[7] family of compact open sets of $\mathscr{S}$, and $U_k \cap F \neq \varnothing$, then $\bigcap (U_k \mid k \in K) \cap F \neq \varnothing$.*

REMARK.   The meaning of condition (S1) is clear. Condition (S2) is a complicated way of ensuring that (C) holds and that Lemma 2 holds for the join-semilattice of compact open sets of $\mathscr{S}(L)$.

PROOF.   To show that (S1) holds, we have to verify that the $r(a), a \in L$, form a base (not only a subbase) for the open sets of $\mathscr{S}(L)$. In other words, for $a,b \in L$, $P \in r(a) \cap r(b)$, we have to find a $c \in L$ with $P \in r(c), r(c) \subseteq r(a) \cap r(b)$. By assumption, $a \notin P$, $b \notin P$; thus, if P is prime, there exists a $c \in L, c \notin P, c \leq a, c \leq b$. Then $P \in r(c), r(c) \subseteq r(a)$, and $r(c) \subseteq r(b)$, as required. To verify (S2) for $\mathscr{S}(L)$, let $F = \mathscr{S}(L) - r(I)$ and $U_k = r(a_k)$. Thus $F = \{P \mid P \supseteq I\}$ and $U_k = \{P \mid a_k \notin P\}$. The assumptions on the a_k mean that $D = \{x \mid x \geq a_k$ for some $k \in K\}$ is a dual ideal; since $U_k \cap F \neq \varnothing$, we have $r(a_k) \nsubseteq r(I)$; that is, $a_k \notin I$, showing that $D \cap I = \varnothing$. Therefore, by Lemma 2, there exists a prime ideal P with $P \supseteq I, P \cap D = \varnothing$. Then $a_k \notin P$, and so $P \in r(a_k)$ for all $k \in K$. Also $P \supseteq I$, thus $P \notin r(I)$, and so $P \in F$, proving that $P \in F \cap \bigcap (U_k \mid k \in K)$, verifying (S2).

Conversely, let $\mathscr{S}$ be a topological space satisfying conditions (S1) and (S2) of the theorem. Let L be the set of compact open sets of $\mathscr{S}$. Obviously, $\varnothing \in L$ and if $A,B \in L$, then $A \cup B \in L$, and thus L is a join-semilattice with zero. Let

$$A \subseteq B_0 \cup B_1 \quad \text{with } A,B_0,B_1 \in L.$$

---

[7] $K \neq \varnothing$ and for $k_0, k_1 \in K$, there exists a $k_2 \in K$ such that $U_{k_2} \subseteq U_{k_0} \cap U_{k_1}$.

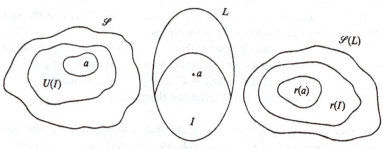

Figure 11.3

Then $A \cap B_i$ is open, and therefore $A \cap B_i = \bigcup (A_j^i \mid j \in J_i)$, $i = 0, 1$, where the A_j^i are compact open sets. Since $A = (A \cap B_0) \cup (A \cap B_1) \subseteq \bigcup (A_j^i \mid j \in J_0 \cup J_1, i = 0, 1)$, by the compactness of A we get $A \subseteq \bigcup (A_j^i \mid j \in \bar{J}_0 \text{ or } j \in \bar{J}_1)$, where $\bar{J}_i$ is a finite subset of J_i, $i = 0, 1$. Set

$$A_i = \bigcup (A_j^i \mid j \in \bar{J}_i), \qquad i = 0, 1.$$

Then $A_0, A_1 \in L$, $A = A_0 \cup A_1$, and $A_0 \subseteq B_0$, $A_1 \subseteq B_1$, showing that L is distributive.

It follows from (S1) that the open sets of $\mathscr{S}$ are uniquely associated with ideals of L: For an ideal I of L, let $U(I) = \bigcup (a \mid a \in I)$ (keep in mind that an $a \in L$ is a subset of $\mathscr{S}$, as illustrated by Figure 11.3). Note that for $a \in L$, $a \in I$ iff $a \subseteq U(I)$.

Now let P be a prime ideal of L, $F = \mathscr{S} - U(P)$, and let $\{U_k \mid k \in K\}$ be the set of all compact open sets of $\mathscr{S}$ that have nonvoid intersections with F. Thus, the U_k are exactly those elements of L that are not in P. Therefore, by the definition of a prime ideal, given $k, l \in K$, there exists $t \in K$ with $U_t \subseteq U_k$, $U_t \subseteq U_l$, proving that F and $\{U_k \mid k \in K\}$ satisfy the hypothesis of condition (S2). By (S2) we conclude that there exists a $p \in F \cap \bigcap (U_k \mid k \in K)$. If $q \in F$, then for every compact open set U with $q \in U$, we have $U \cap F \neq \varnothing$; thus $p \in U$, proving that $\overline{\{p\}} = F$. Note that $\mathscr{S}$ is T_0; therefore p is unique.

Conversely, if $p \in \mathscr{S}$, let $I = \{a \mid a \in L, a \subseteq \mathscr{S} - \overline{\{p\}}\}$. Then I is an ideal of L, and $\mathscr{S} - \overline{\{p\}} = U(I)$. We claim that I is prime. Indeed, if $U, V \in L$, $U \notin I$, $V \notin I$, then $U \cap \overline{\{p\}} \neq \varnothing$, $V \cap \overline{\{p\}} \neq \varnothing$, and therefore $p \in U$, $p \in V$. Thus, $p \in U \cap V$ and so $U \cap V \nsubseteq U(I)$. By (S1) there exists a $W \in L$ with $W \subseteq U \cap V$, $W \nsubseteq U(I)$. Therefore $W \notin I$, and so I is prime.

Summing up, the map $\varphi: P \to p$ is one-to-one and onto between $\mathscr{S}(L)$

and $\mathscr{S}$. To show that φ is a homeomorphism, it suffices to show that U is open in $\mathscr{S}(L)$ iff $U\varphi$ is open in $\mathscr{S}$. Since a typical open set in $\mathscr{S}(L)$ is of the form $r(I)$ $(I \in I(L))$, and an open set of $\mathscr{S}$ is of the form $U(I)$, we need only prove that $r(I)\varphi = U(I)$ and $U(I)\varphi^{-1} = r(I)$—in other words, that $P \in r(I)$ iff $(P\varphi =) p \in U(I)$. Indeed, $P \in r(I)$ means that $P \nsupseteq I$, which is equivalent to $U(P) \nsupseteq U(I)$; this, in turn, is the same as $U(I) \cap (\mathscr{S} - U(P)) \neq \varnothing$. Since $\mathscr{S} - U(P) = \overline{\{p\}}$ with $p = P\varphi$, the last condition means that $U(I) \cap \overline{\{p\}} \neq \varnothing$, which holds iff $p \in U(I)$. (Indeed, if $p \notin U(I)$, then $U(I) \subseteq U(P)$, and so $U(I) \cap \overline{\{p\}} = \varnothing$.) ●

COROLLARY 9 (M. H. Stone [1937]). *The Stone space of a distributive lattice is characterized by* (S1), (S2), *and*

(S3) *The intersection of two compact open sets is compact.*

COROLLARY 10 (M. H. Stone [1936]). *The Stone space $\mathscr{S}$ of a Boolean lattice (called a* Boolean space*) can be characterized as a compact Hausdorff space in which the closed open* (clopen) *sets form a base for open sets. (In other words, $\mathscr{S}$ is* totally disconnected.*)*

PROOF. Let $\mathscr{S} = \mathscr{S}(B)$, where B is a Boolean lattice. Then $\mathscr{S} = r(1)$, and thus $\mathscr{S}$ is compact. Let $P, Q \in \mathscr{S}$ and $P \neq Q$ and take $a \in P - Q$. Then $Q \in r(a)$, $P \in r(a')$; therefore, every pair of elements of $\mathscr{S}$ can be separated by clopen sets, verifying that $\mathscr{S}$ is Hausdorff. Obviously, $\mathscr{S}$ is totally disconnected. Conversely, let $\mathscr{S}$ be compact, Hausdorff, and totally disconnected. Then (S1) is obvious. (S2) follows from the observation that F and the U_i, $i \in I$, are now closed sets having the finite intersection property; therefore, by compactness, they have an element in common. The compact open sets of $\mathscr{S}$ form a Boolean lattice B, and thus $\mathscr{S}$ is homeomorphic to $\mathscr{S}(B)$. ●

As an interesting application we prove:

THEOREM 11. *Let B be an infinite Boolean lattice. Then $|\mathscr{P}(B)| \geq |B|$.*

PROOF. Let $\mathscr{S}$ be a totally disconnected compact Hausdorff space. For $a, b \in \mathscr{S}$ with $a \neq b$, fix a pair of clopen sets $U_{a,b}$ and $U_{b,a}$ such that $a \in U_{a,b}$, $b \in U_{b,a}$, and $U_{a,b} \cap U_{b,a} = \varnothing$. Now let U be clopen and $a \in U$. Then $\mathscr{S} - U \subseteq \bigcup (U_{b,a} \mid b \in \mathscr{S} - U)$, and so, by the compactness of $\mathscr{S} - U$, $\mathscr{S} - U \subseteq \bigcup (U_{b,a} \mid b \in B)$ for some finite $B \subseteq \mathscr{S} - U$. Then

$V_a = \bigcap (U_{a,b} \mid b \in B)$ is open and $a \in V_a \subseteq U$. Thus, $U \subseteq \bigcup (V_a \mid a \in U)$, so by the compactness of U, $U \subseteq \bigcup (V_a \mid a \in A)$ for some finite $A \subseteq U$. Therefore, $U = \bigcup (V_a \mid a \in A)$. Thus every clopen set is a finite union of finite intersections of $U_{a,b}$, and so there are no more clopen sets than there are finite sequences of elements of $\mathscr{S}$; this cardinality is $|\mathscr{S}|$ provided that $|\mathscr{S}|$ is infinite. ●

It might be illuminating to compare this to an algebraic proof; see exercise 36.

Theorem 8 and its corollaries provide topological representations for distributive join-semilattices, distributive lattices, and Boolean lattices, respectively. It is possible to give topological representation for homomorphism. We do it here only for $\{0, 1\}$-homomorphisms of bounded distributive lattices.

LEMMA 12. *Let L_0 and L_1 be bounded distributive lattices and let φ be a $\{0, 1\}$-homomorphism of L_0 into L_1. Then*

$$\mathscr{S}(\varphi): P \to P\varphi^{-1}$$

maps $\mathscr{S}(L_1)$ into $\mathscr{S}(L_0)$; $\mathscr{S}(\varphi)$ is a continuous function with the property that if U is compact open in $\mathscr{S}(L_0)$, then $U(\mathscr{S}(\varphi))^{-1}$ is compact in $\mathscr{S}(L_1)$. Conversely, if $\psi: \mathscr{S}(L_1) \to \mathscr{S}(L_0)$ has these properties, then $\psi = \mathscr{S}(\varphi)$ for exactly one $\varphi: L_0 \to L_1$.

PROOF. If $U = r(a)$, $a \in L_0$, then

$$U\mathscr{S}(\varphi)^{-1} = \{P \mid P \in \mathscr{S}(L_1), P\varphi^{-1} \in r(a)\}$$

$$= \{P \mid P \in \mathscr{S}(L_1), a \notin P\varphi^{-1}\}$$

$$= \{P \mid P \in \mathscr{S}(L_1), a\varphi \notin P\} = r(a\varphi),$$

and so $\mathscr{S}(\varphi)$ is continuous, having the desired property.

Conversely, if such a ψ is given and $U = r(a)$, $a \in L_0$, then $U\psi^{-1}$ is compact open, and so $U\psi^{-1} = r(b)$ for a unique $b \in L_1$. The map $\varphi: a \to b$ is a $\{0, 1\}$-homomorphism, and $\psi = \mathscr{S}(\varphi)$. ●

The following interpretation of (S2) will be useful. Let $\mathscr{S}$ be a topological space. The *Booleanization of $\mathscr{S}$* is a topological space $\mathscr{S}^B$ on $\mathscr{S}$ that has the compact open sets of $\mathscr{S}$ and their complements as a subbase for open sets.

LEMMA 13. *The compact topological space $\mathscr{S}$ satisfies* (S1), (S2), *and* (S3) *iff $\mathscr{S}^B$ is a Boolean space.*

PROOF. Let $\mathscr{S}$ satisfy (S1), (S2), and (S3). Then $\mathscr{S}^B$ is obviously Hausdorff and totally disconnected. To verify the compactness of $\mathscr{S}^B$, let $\mathscr{F}_0$ be a collection of compact open sets of $\mathscr{S}$ and let $\mathscr{F}_1$ be a collection of complements of compact open sets of $\mathscr{S}$ such that in $\mathscr{F} = \mathscr{F}_0 \cup \mathscr{F}_1$ no finite intersection is void. Because of (S3) we can assume that $\mathscr{F}_0$ is closed under finite intersection. Since members of $\mathscr{F}_1$ are closed in $\mathscr{S}$ and $\mathscr{S}$ is compact, $\bigcap (X \mid X \in \mathscr{F}_1) = F$ is nonvoid. Also, for any $U \in \mathscr{F}_0$, and $X \in \mathscr{F}_1$, $U \cap X$ is closed in U, and thus $U \cap F = \bigcap (U \cap X \mid X \in \mathscr{F}_1) \neq \varnothing$. Applying (S2) to F and $\mathscr{F}_0$, we conclude that $\bigcap \mathscr{F} \neq \varnothing$, which, by Alexander's Theorem (see exercise 14), proves compactness.

Conversely, if $\mathscr{S}^B$ is Boolean, the compact open sets of $\mathscr{S}^B$ form a Boolean lattice L. We can easily verify that the compact open sets of $\mathscr{S}$ form a sublattice L_1 of L. Thus L_1 is a distributive lattice, and so, by Corollary 9, $\mathscr{S} = \mathscr{S}(L_1)$ satisfies (S1), (S2), and (S3). ●

Exercises

The following exercises review the basic concepts and theorems of topology that are utilized in Sections 11 and 12.

1. A *topological space* is a set A and a collection T of subsets of A closed under finite intersections and arbitrary unions; a member of T is called an *open set*. Call a set *closed* if its complement is open. Characterize the family of closed sets.

2. A family of nonvoid sets $B \subseteq T$ is a *base* for open sets if every open set is a union of members of B. Show that for a set A, $B \subseteq P(A)$ is a base for open sets of some topological space defined on A iff $\bigcup B = A$, and for $X, Y \in B$ and $p \in X \cap Y$, there exists a $Z \in B$ with $p \in Z$, $Z \subseteq X$, and $Z \subseteq Y$.

3. A family of nonvoid sets $C \subseteq P(A)$ is a *subbase* for open sets if the finite intersections of members of C form a base for open sets. Show that $C \subseteq P(A)$ is a subbase of some topology defined on A iff $\bigcup C = A$.

4. Let A be a topological space and let $X \subseteq A$. Then there exists a smallest closed set $\bar{X}$ containing X; $\bar{X}$ is the *closure* of X. Show that for $X, Y \subseteq A$, $X \subseteq Y$ implies that $\bar{X} \subseteq \bar{Y}$; $X \subseteq \bar{X}$; $\overline{X \cup Y} = \bar{X} \cup \bar{Y}$; and $\bar{\bar{X}} = \bar{X}$.

5. Prove that the four properties of $\bar{X}$ given in exercise 4 characterize it.

6. Show that $a \in \bar{X}$ iff every open set (in a given base) containing a has a nonvoid intersection with X.

7. A space A is a T_0-*space* if $\overline{\{x\}} = \overline{\{y\}}$ implies that $x = y$. Show that A is a T_0-space iff, for every $x,y \in A$, $x \neq y$, there exists an open set (in a given base) containing exactly one of x and y.

8. A is a T_1-*space* if $\overline{\{x\}} = \{x\}$ for all $x \in A$. A T_1-space is a T_0-space. Show that A is a T_1-space iff, for $x,y \in A$, $x \neq y$, there exists an open set (in a given base) containing x but not y.

9. Let A and B be topological spaces and $f: A \to B$. Then f is called *continuous* if, for every open set U of B, $f^{-1}(U)$ is open in A. f is a *homeomorphism* if f is one-to-one and onto and if both f and f^{-1} are continuous. Show that continuity can be checked by considering only those $f^{-1}(U)$ where U belongs to a given subbase.

10. Show that $f: A \to B$ is continuous iff $f(\bar{X}) \subseteq \overline{f(X)}$ for all $X \subseteq A$.

11. A subset X of a topological space A is *compact* if $X \subseteq \bigcup (U_i \mid U_i$ open, $i \in I)$ implies that $X \subseteq \bigcup (U_i \mid i \in I_1)$ for some finite $I_1 \subseteq I$.
   The space A is *compact* if $X = A$ is compact. Show that A is compact iff, for every family F of closed sets, if $\bigcap F_1 \neq \varnothing$ for all finite $F_1 \subseteq F$, then $\bigcap F \neq \varnothing$.

12. Let A be a compact topological space and let X be a closed set in A. Show that X is compact.

13. Prove that a space A is compact iff, in the lattice of closed sets of A, every maximal dual ideal is principal.

*14. Show that a T_1-space A is compact iff it has a subbase C of closed sets (that is, $\{A - X \mid X \in C\}$ is a subbase for open sets) with the property: If $\bigcap D = \varnothing$ for some $D \subseteq C$, then $\bigcap D_1 = \varnothing$ for some finite $D_1 \subseteq D$ (J. W. Alexander's Theorem [1939]).

15. Let $A_i, i \in I$, be topological spaces and set $A = \prod (A_i \mid i \in I)$. For $U \subseteq A_i$, set $E(U) = \{f \mid f \in A, f(i) \in U\}$. The *product topology* on A is the topology determined by taking all the sets $E(U)$ as a subbase for open sets, where U ranges over all open sets of A_i for all $i \in I$. Show that the projection map $e_i: f \to f_i$ is a continuous map of A onto A_i.

16. Show that if the A_i are T_0-spaces (T_1-spaces), so is $A = \prod (A_i \mid i \in I)$.

17. A map $f: A \to B$ is *open* if $f(U)$ is open in B for every open $U \subseteq A$. Show that the projection maps (see exercise 15) are open.

18. Prove that a function $f: B \to \prod A_i$ is continuous iff, for each $i \in I$, $e_i \cdot f: B \to A_i$ is continuous.

19. A space A is a *Hausdorff space* (T_2-*space*) if, for $x,y \in A$ with $x \neq y$, there exist open sets U, V such that $x \in U$, $y \in V$, $U \cap V = \varnothing$. Show that:
   (i) A is Hausdorff iff $\Delta = \{\langle x, x \rangle \mid x \in A\}$ is closed in $A \times A$.
   (ii) A compact subset of a T_2-space is closed.

20. Prove that a product of Hausdorff spaces is a Hausdorff space.

21. Show that a product of compact spaces is compact (Tihonov's Theorem). (Hint: use exercise 14.)

22. A space A is *totally disconnected* if, for $x,y \in A$, $x \neq y$, there exists a

closed open set U with $x \in U$, $y \notin U$. Show that the product of totally disconnected sets is totally disconnected.

Following are some exercises for Section 11.

23. Let I and J be ideals of a join-semilattice. Verify that

$$I \vee J = \{t \mid t \le i \vee j, i \in I, j \in J\}.$$

24. Show that for a join-semilattice L, $I(L)$ is a lattice iff any two elements of L have a lower bound.

25. Give a detailed proof of Lemma 2.

26. Prove that every join-semilattice can be embedded in a Boolean lattice (considered as a join-semilattice).

27. Show that a finite distributive join-semilattice is a distributive lattice.

28. Let L be a join-semilattice and let Θ be an equivalence relation on L having the Substitution Property for join. Then L/Θ is also a join-semilattice. Show that the distributivity of L does not imply the distributivity of L/Θ.

29. Let F be a free join-semilattice; let F_0 be F with a new 0 added. Show that F_0 is a distributive join-semilattice.

30. Let φ be a join-homomorphism of the join semilattice F_0 into the join-semilattice F_1. Show that, for distributive join-semilattices F_0, F_1, the proper homomorphism concept is the one requiring that if P is a prime ideal of F_1, then $P\varphi^{-1}$ is a prime ideal of F_0.

31. Show that there is no "free distributive join-semilattice" with the homomorphism concept of exercise 30.

32. Does Theorem 7.22 generalize to distributive join-semilattices?

33. Characterize the Stone spaces of finite Boolean lattices and of finite chains.

34. Let $\mathscr{S}_0$ and $\mathscr{S}_1$ be disjoint topological spaces; let $\mathscr{S} = \mathscr{S}_0 \cup \mathscr{S}_1$ and call $U \subseteq \mathscr{S}$ open if $U \cap \mathscr{S}_0$ and $U \cap \mathscr{S}_1$ are open. Show that if $\mathscr{S}_0$ and $\mathscr{S}_1$ are Stone spaces, then so is $\mathscr{S}$.

35. If, in exercise 34, $\mathscr{S}_i = \mathscr{S}(L_i)$, $i = 0, 1$, then $\mathscr{S} = \mathscr{S}(L_0 \times L_1)$.

36. To prove Theorem 11, pick an element $a(P, Q) \in P - Q$ for all $P, Q \in \mathscr{P}(B)$ with $P \ne Q$. Show that the $a(P, Q)$ R-generate all of B. (Corollary 8.7 or Lemma 13.3 can be used.)

37. In Lemma 12, characterize $\mathscr{S}(\varphi)$ for one-to-one and for onto φ.

38. Determine the connection between the Stone space of a lattice and the Stone space of a sublattice.

39. Call the Stone space of a generalized Boolean lattice a *generalized Boolean space*; characterize it. (Compactness of $\mathscr{S}$ should be replaced by *local compactness*: For every $p \in \mathscr{S}$ there exists an open set U with $p \in U$ and a set V with $U \subseteq V$, such that V is compact.)

40. Show that the product of (generalized) Boolean spaces is (generalized) Boolean.

41. Call the join-semilattice L *modular* if, for $a,b,c \in L$, $a \le b$ and $b \le a \vee c$

imply the existence of $c_1 \in L$ with $c_1 \le c, b = a \lor c_1$. Show that a distributive join-semilattice is modular.

42. Show that Lemma 1 remains valid if all occurrences of the word "distributive" are replaced by the word "modular."

43. Show that the set of all finitely generated normal subgroups of a group (and also the finitely generated ideals of a ring) form a modular join-semilattice.

44. The lattice of congruence relations of a join-semilattice L is distributive iff any pair of elements with a lower bound are comparable (D. Papert [1964], R. A. Dean and R. H. Oehmke [1964]).

12. Free Distributive Product

Let L_i, $i \in I$, be pairwise disjoint distributive lattices. Then $Q = \bigcup (L_i \mid i \in I)$ is a partial lattice. A free lattice generated by Q over the class **D** of all distributive lattices is called a *free distributive product* of the L_i, $i \in I$. To prove the existence of free distributive products, it suffices by Theorem 5.24 to show that there exists a distributive lattice L containing Q as a relative sublattice. This is easily done: Let L be the direct product of the $L_i \cup \{0\}$, $i \in I$, where 0 is a new zero element of L_i. Identify $x \in L_i$ with $f \in L$ defined by $f(i) = x, f(j) = 0$ for $j \ne i$. Then Q becomes a relative sublattice of L.

An equivalent definition is:

DEFINITION 1.   *Let* **K** *be a class of lattices and let* L_i, $i \in I$, *be lattices in* **K**. *A lattice* L *in* **K** *is called a* free **K**-product *of the* L_i, $i \in I$, *if every* L_i *has an embedding* ϵ_i *into* L *such that*

(i) L *is generated by* $\bigcup (L_i \epsilon_i \mid i \in I)$.

(ii) *If* K *is any lattice in* **K** *and* φ_i *is a homomorphism of* L_i *into* K, *for* $i \in I$, *then there exists a homomorphism* φ *of* L *into* K *satisfying* $\varphi_i = \epsilon_i \varphi$ *for all* $i \in I$ (*see Figure 12.1*).

For distributive lattices, this is equivalent to the first definition (see exercises at the end of this section). In most cases we will assume that each L_i is a sublattice of L and that ϵ_i is the inclusion map; then (ii) will simply state that the φ_i have a common extension. Note that in all the cases we shall consider, (i) can be replaced by the requirement that the φ in (ii) be unique.

If, in Definition 1, **K** is the class of bounded distributive lattices and all

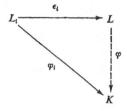

Figure 12.1

homomorphisms are assumed to be $\{0, 1\}$-homomorphisms, we get the concept of a *free $\{0, 1\}$-distributive product*.

Our first result is the existence and description of a free $\{0, 1\}$-distributive product of a family of bounded distributive lattices.

THEOREM 2 (A. Nerode [1959]).   *Let L_i, $i \in I$, be distributive lattices with 0 and 1. Let $\mathscr{S} = \prod (\mathscr{S}(L_i) \mid i \in I)$ (see exercise 11.15). Then $\mathscr{S}$ is a Stone space, and thus $\mathscr{S} \cong \mathscr{S}(L)$ for some distributive lattice L. Any such lattice L is a free $\{0, 1\}$-distributive product of the L_i, $i \in I$.*

The proof of Theorem 2 will be preceded by two lemmas. In these two lemmas a *Stone space* is a topological space satisfying (S1), (S2), and (S3) of Section 11.

LEMMA 3.   *Let $\mathscr{S}_i$, $i \in I$, be compact Stone spaces. Then*

$$\prod (\mathscr{S}_i^B \mid i \in I) = (\prod (\mathscr{S}_i \mid i \in I))^B.$$

(The notation $\mathscr{S}^B$ was introduced in Section 11.)

PROOF.   For $U \subseteq \mathscr{S}_j$ let $E(U) = \{f \mid f \in \prod \mathscr{S}_i, f(j) \in U\}$ (see exercise 11.15). The compact open sets form a base for open sets in $\mathscr{S}_j$; therefore, $\{E(U) \mid U$ compact open in some $\mathscr{S}_j\}$ is a subbase for open sets in $\prod (\mathscr{S}_i \mid i \in I)$. Note that all the sets $E(U)$ are compact open in $\prod \mathscr{S}_i$; therefore, $V \subseteq \prod \mathscr{S}_i$ is compact open iff it is a finite union of finite intersections of some of the $E(U)$. Consequently, declaring the complements of compact open sets to be open (in forming $(\prod \mathscr{S}_i)^B$) is equivalent to making the complements of the sets $E(U)$ open. But the complement of $E(U)$ is $E(\mathscr{S}_i - U)$, and $\mathscr{S}_i - U$ is a typical open set of $\mathscr{S}_i^B$. Thus $\prod \mathscr{S}_i^B$ and $(\prod \mathscr{S}_i)^B$ have the same topology.   ●

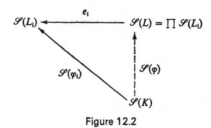

Figure 12.2

LEMMA 4.   *The product of compact Stone spaces is again a compact Stone space.*

PROOF.   Let $\mathscr{S}_i$, $i \in I$, be Stone spaces. Then $\mathscr{S} = \prod \mathscr{S}_i$ is T_0 and compact (exercises 11.16 and 11.21). Since $\mathscr{S}_i^B$ is Boolean (Lemma 11.13), so is $\prod \mathscr{S}_i^B$ (exercises 11.20, 11.21, and 11.22). By Lemma 3, $\mathscr{S}^B = \prod \mathscr{S}_i^B$, and thus $\mathscr{S}^B$ is Boolean. Therefore, by Lemma 11.13, $\mathscr{S}$ is a Stone space.   ●

PROOF OF THEOREM 2.   Let e_i be the ith projection $(e_i : \mathscr{S}(L) \to \mathscr{S}(L_i)$, $fe_i = f(i))$. By Lemma 11.2, there is a unique $\{0, 1\}$-homomorphism $\epsilon_i : L_i \to L$ satisfying $\mathscr{S}(\epsilon_i) = e_i$. It is easy to visualize ϵ_i; think of the elements of L_i as compact open sets of $\mathscr{S}_i$; then $U\epsilon_i = E(U) = Ue_i^{-1}$. It is obvious from this that ϵ_i is an embedding.

By applying $\mathscr{S}$ to Figure 12.1, we obtain Figure 12.2. Thus the method of defining $\mathscr{S}(\varphi)$ is obvious. For $x \in \mathscr{S}(K)$, $x\mathscr{S}(\varphi)$ is a member of $\prod \mathscr{S}(L_i)$, and $x\mathscr{S}(\varphi)(i) = x\mathscr{S}(\varphi_i)$ for $i \in I$. To show that this correspondence is indeed of the form $\mathscr{S}(\varphi)$ for some homomorphism $\varphi : L \to K$, we have to verify that (a) $\mathscr{S}(\varphi)$ is continuous (this statement follows from exercise 11.18), and that (b) if V is compact open in $\mathscr{S}(L)$, then $V\mathscr{S}(\varphi)^{-1}$ is compact open in $\mathscr{S}(K)$. It is enough to verify (b) for $V = E(U)$ where U is compact open in some $\mathscr{S}(L_i)$. Then

$$V\mathscr{S}(\varphi)^{-1} = E(U)\mathscr{S}(\varphi)^{-1} = Ue_i^{-1}\mathscr{S}(\varphi)^{-1} = U(\mathscr{S}(\varphi)e_i)^{-1} = U\mathscr{S}(\varphi_i)^{-1},$$

and therefore $V\mathscr{S}(\varphi)^{-1}$ is compact open since $\mathscr{S}(\varphi_i)$ satisfies the condition of Lemma 11.12.   ●

This result not only establishes the existence of free $\{0, 1\}$-distributive products but gives a useful description of them. Some applications will be given in the exercises at the end of this section and other applications will be shown in Chapter 3.

Let L be a free $\{0, 1\}$-distributive product of the L_i, $i \in I$, and let us assume for the rest of this section that $L_i \subseteq L$. Thus $L = [\bigcup L_i]$, and so every $a \in L$ can be written in the form

$$a = \bigvee (\bigwedge X \mid X \in J)$$

(and dually) where every $X \in J$ is a finite subset of $\bigcup L_i$, and J is finite. (This is obvious directly, but it also follows from Lemma 4.3 and the representation of distributive lattice polynomials given in the proof of Theorem 8.1.) Such representations are not unique. Therefore, the question arises as to how we can determine whether two such expressions represent the same element. This is called the word problem, and we will give a solution to it based on a characterization of free products. (All the subsequent results of this section are from G. Grätzer and H. Lakser [1969a].)

THEOREM 5.   *Let L be a distributive lattice with 0 and 1, let the L_i be $\{0, 1\}$-sublattices of L for $i \in I$, and let $L = [\bigcup L_i]$. Then L is a free $\{0, 1\}$-distributive product of the L_i iff, for finite nonvoid $I_0, I_1 \subseteq I$, for $x_i \in L_i$, $i \in I_0$, for $y_j \in L_j$, $j \in I_1$, satisfying $x_k, y_k \notin \{0, 1\}$ in L_k, for all $k \in I$,*

$$\bigwedge (x_i \mid i \in I_0) \leq \bigvee (y_j \mid j \in I_1)$$

implies the existence of $i \in I_0 \cap I_1$ with $x_i \leq y_i$.

REMARK.   If there is an $i \in I_0 \cap I_1$ with $x_i \leq y_i$, then $\bigwedge (x_i \mid i \in I_0) \leq \bigvee (y_j \mid j \in I_1)$. Thus the condition should be interpreted as requiring that $\bigwedge (x_i \mid i \in I_0) \leq \bigvee (y_j \mid j \in I_1)$ holds only in the trivial case. This result is a direct generalization of Theorem 8.3.

PROOF.   Let L be a free $\{0, 1\}$-distributive product of the L_i, $i \in I$, $\bigwedge (x_i \mid i \in I_0) \leq \bigvee (y_j \mid j \in I_1)$ and let $x_i \not\leq y_i$ for every $i \in I_0 \cap I_1$. For every $i \in I$ we choose a prime ideal P_i of L_i as follows: If $i \in I_0 \cap I_1$, then $x_i \not\leq y_i$, so we can choose P_i with $x_i \notin P_i$ and $y_i \in P_i$; for $i \in I_0 - I_1$, we assume that $x_i \notin P_i$; for $i \in I_1 - I_0$, we assume that $y_i \in P_i$. Now we define homomorphisms $\varphi_i : L_i \to \mathfrak{C}_2$ by $x\varphi_i = 0$ for $x \in P_i$, $x\varphi_i = 1$ for $x \notin P_i$. Since L is a free product, there is a common extension $\varphi : L \to \mathfrak{C}_2$ of the φ_i. But this is impossible because $\bigwedge (x_i \mid i \in I_0)\varphi = \bigwedge (x_i\varphi \mid i \in I_0) = \bigwedge 1 = 1$, and $\bigvee (y_j \mid j \in I_1)\varphi = \bigvee 0 = 0$, contradicting $\bigwedge x_i \leq \bigvee y_j$.

Now let L satisfy the condition of the theorem. Let L^* be a free $\{0, 1\}$-distributive product of the L_i, $i \in I$ (we assume $L_i \subseteq L^*$), let φ_i be the identity map on L_i as a map from L_i into L, and let φ be the common extension of the φ_i, $i \in I$. We claim that φ is an isomorphism. Since $L^*\varphi$ includes all $L_i(=L_i\varphi)$ and the L_i generate L, φ is onto. So to show that φ is an isomorphism we have only to verify that for $a,b \in L^* - \{0, 1\}$,

$$a\varphi \leq b\varphi \quad \text{implies that} \quad a \leq b.$$

Let

$$a = \bigvee (\bigwedge (x_i \mid i \in I_j) \mid j \in J)$$

and

$$b = \bigwedge (\bigvee (y_k \mid k \in K_m) \mid m \in M),$$

where I_j and K_m are finite nonvoid subsets of I, $x_i \in L_i$, $y_k \in L_k$, and J and M are finite and nonvoid. If $a\varphi \leq b\varphi$, then for each $j \in J$ and $m \in M$,

$$\bigwedge (x_i \mid i \in I_j) \leq \bigvee (y_k \mid k \in K_m)$$

in L; thus, by assumption, for some $i \in I_j \cap K_m$, $x_i \leq y_i$. This implies that

$$\bigwedge (x_i \mid i \in I_j) \leq \bigvee (y_k \mid k \in K_m)$$

in L^*; therefore $a \leq b$. ●

As an immediate corollary, we get the following result of B. Jónsson [1961] (see also Section 13).

COROLLARY 6. *Let K_i be a $\{0, 1\}$-sublattice of the distributive lattice L_i with 0 and 1, for $i \in I$. Let L be a free $\{0, 1\}$-distributive product of the L_i (assume that $L_i \subseteq L$) and $K = [\bigcup (K_i \mid i \in I)]$. Then K is a free $\{0, 1\}$-distributive product of the K_i, $i \in I$.*

Another consequence is:

COROLLARY 7. *With notation as in Theorem 5, let I_0, I_1 be finite nonvoid subsets of I. Let $x_i \in L_i - \{0, 1\}$ for $i \in I_0$ and let $y_j \in L_j - \{0, 1\}$ for $j \in I_1$. Then*

$$\bigwedge (x_i \mid i \in I_0) \leq \bigwedge (y_j \mid j \in I_1),$$

iff $I_0 \supseteq I_1$ and for $i \in I_1$, $x_i \leq y_i$.

PROOF. The "if" part is trivial. Conversely, if $\bigwedge x_i \leq \bigwedge y_j$, then $\bigwedge x_i \leq y_j$ for every $j \in I_1$. Applying Theorem 5, we conclude that $I_0 \cap \{j\}$ is nonvoid, that is, that $j \in I_0$. Thus $I_0 \supseteq I_1$, and $x_i \leq y_i$ for $i \in I_1$. ●

Every element a of a free $\{0, 1\}$-distributive product L of the L_i has many representations of the form $\bigvee (\bigwedge (x_i \mid i \in I_j) \mid j \in J)$. Using Theorem 5, we can decide whether two such expressions represent the same element. Thus Theorem 5 solves the word problem. It is desirable, however, to select one of these expressions (which will be called *normal form*). We start with a few definitions and notations.

Let L be a free $\{0, 1\}$-distributive product of L_i, $i \in I$, and let

$$Q = \bigcup (L_i - \{0, 1\} \mid i \in I) \cup \{0, 1\}.$$

We identify $0 \in Q$ with $0 \in L_i$ for each $i \in I$, and similarly for $1 \in Q$. A finite nonempty subset $X \subseteq Q$ is said to be *reduced* if $|X \cap L_i| \leq 1$ for all $i \in I$. It should be noted that if X is reduced and $0 \in X$, then $X = \{0\}$, and dually. If $X \subseteq Q$ is finite and nonempty, we can define a reduced subset $X^\wedge$ of Q, the $\wedge$-*reduct* of X, by the conditions:

    (i) If $I' = \{i \in I \mid X \cap L_i \neq \varnothing\}$, then

$$X^\wedge = \{\bigwedge (X \cap L_i) \mid i \in I'\},$$

    provided that $\bigwedge (X \cap L_i) \neq 0$ for all $i \in I'$.

    (ii) If there is an $i \in I'$ such that $\bigwedge (X \cap L_i) = 0$, then $X^\wedge = \{0\}$. The $\vee$-*reduct* of X, denoted $X^\vee$, is defined in the dual manner. We note that, in L, $\bigwedge X = \bigwedge (X^\wedge)$ and $\bigvee X = \bigvee (X^\vee)$.

DEFINITION 8. *A finite nonempty family J of finite reduced subsets of Q is said to be a $\bigvee$-representation of $a \in L$ if*

$$a = \bigvee (\bigwedge X \mid X \in J).$$

The family J is said to be a $\bigwedge$-representation of $a \in L$ if

$$a = \bigwedge (\bigvee X \mid X \in J).$$

Obviously, every element a of L has both $\bigvee$- and $\bigwedge$-representations. Given a $\bigwedge$-representation J of an element $a \in L$, we can write, using distributivity (see exercise 4.11),

$$a = \bigvee (\bigwedge F(J) \mid F \in \mathscr{C}(J)),$$

where $\mathscr{C}(J)$ denotes the set of *choice functions* on J—that is, the set of functions $F: J \to \bigcup J$ such that $F(X) \in X$ for each $X \in J$. As a result of our previous discussion, we find that $a = \bigvee (\bigwedge (F(J)^{\wedge}) \mid F \in \mathscr{C}(J))$. Since the set $\mathscr{C}(J)$ is finite, we can consider a subset $\mathscr{C}_{\mathrm{red}}(J) \subseteq \mathscr{C}(J)$, the set of *reduced* choice functions such that the set

$$\{\bigwedge (F(J)^{\wedge}) \mid F \in \mathscr{C}_{\mathrm{red}}(J)\}$$

is the set of all maximal elements of the set

$$\{\bigwedge (F(J)^{\wedge}) \mid F \in \mathscr{C}(J)\}.$$

Thus the family $\{F(J)^{\wedge} \mid F \in \mathscr{C}_{\mathrm{red}}(J)\}$ is a $\bigvee$-representation of a; it is said to be a *normal $\bigvee$-representation* of a. There is, of course, dually a *normal $\bigwedge$-representation* of a.

THEOREM 9.
(i) *Each $a \in L$ has a normal $\bigvee$-representation.*
(ii) *Let $a, b \in L$ and let J_1 be a $\bigvee$-representation of a and J_2 a normal $\bigvee$-representation of b. Then $a \le b$ iff the following condition holds:*
   (R) *For each $X \in J_1$ there is a $Y \in J_2$ such that for each $y \in Y$ there is an $x \in X$ satisfying $x \le y$.*

PROOF. Since

$$a = \bigvee (\bigwedge X \mid X \in J_1)$$

and

$$b = \bigvee (\bigwedge Y \mid Y \in J_2),$$

condition (R) is clearly sufficient for $a \le b$.

   Now let $a \le b$ and let K be a $\bigwedge$-representation of b such that

$$J_2 = \{F(K)^{\wedge} \mid F \in \mathscr{C}_{\mathrm{red}}(K)\}.$$

Therefore,

$$\bigvee (\bigwedge X \mid X \in J_1) \le \bigwedge (\bigvee Z \mid Z \in K).$$

Thus if $X \in J_1$, then $\bigwedge X \le \bigvee Z$ for each $Z \in K$. Since both X and Z are reduced for each $Z \in K$, we conclude, by Theorem 5, that for each $Z \in K$ there is an element $G(Z) \in Z$ such that $\bigwedge X \le G(Z)$. Thus

$$\bigwedge X \le \bigwedge (G(Z) \mid Z \in K) = \bigwedge (G(K)^{\wedge}),$$

where, clearly, $G \in \mathscr{C}(K)$. By the definition of $\mathscr{C}_{red}(K)$, there is an $F \in \mathscr{C}_{red}(K)$ such that $\bigwedge (G(K)^\wedge) \leq \bigwedge (F(K)^\wedge)$. The rest of the condition follows by Corollary 7. ●

Since in a normal $\bigvee$-representation J, the elements of $\{\bigwedge X \mid X \in J\}$ are mutually incomparable, we conclude:

COROLLARY 10.   *The normal $\bigvee$-representation of any element of L is uniquely defined.*

To show how normal representations can be used to investigate the structure of free $\{0, 1\}$-distributive products, we prove:

THEOREM 11.   *Let the L_i, $i \in I$, be distributive lattices with 0 and 1 and let L be a free $\{0, 1\}$-distributive product of L_i, $i \in I$. If all the L_i satisfy the Countable Chain Condition (any chain in L_i is of cardinality $\leq \aleph_0$), then L satisfies the Countable Chain Condition.*

In the proof we will need:

LEMMA 12.   *Let Λ be a chain and let $\mathscr{H} = (H_\lambda \mid \lambda \in \Lambda)$ be a family of finite nonvoid sets. For each pair λ, $\mu \in \Lambda$ such that $\lambda \leq \mu$, let there be a relation $\varphi_{\lambda\mu} \subseteq H_\lambda \times H_\mu$ satisfying these three conditions:*

(i)   *For every $x \in H_\lambda$ there is a $y \in H_\mu$ with $\langle x, y \rangle \in \varphi_{\lambda\mu}$, whenever $\lambda \leq \mu$.*

(ii)   *$\varphi_{\lambda\lambda}$ is equality for all $\lambda \in \Lambda$.*

(iii)   *If $\lambda \leq \mu \leq \nu$, then $\langle x, z \rangle \in \varphi_{\lambda\nu}$ whenever there is a y with $\langle x, y \rangle \in \varphi_{\lambda\mu}$ and $\langle y, z \rangle \in \varphi_{\mu\nu}$.*

Then there is a family $(x_\lambda \mid x_\lambda \in H_\lambda, \lambda \in \Lambda)$ such that $\langle x_\lambda, x_\mu \rangle \in \varphi_{\lambda\mu}$ if $\lambda \leq \mu$.

PROOF.   Consider H_λ to be a discrete topological space. (A topological space $\mathscr{S}$ is *discrete* if every subset of $\mathscr{S}$ is open.) Then H_λ is a compact Hausdorff space, and therefore (by exercise 11.21) so is $H = \prod (H_\lambda \mid \lambda \in \Lambda)$. For $\lambda, \mu \in \Lambda$, $\lambda \leq \mu$, let $S(\lambda, \mu)$ denote the set of all $f \in H$ satisfying $\langle f(\lambda), f(\mu) \rangle \in \varphi_{\lambda\mu}$. Note that $S(\lambda, \mu) \neq \varnothing$ by (i). If $f \notin S(\lambda, \mu)$, then $\{g \mid g(\lambda) = f(\lambda), g(\mu) = f(\mu)\}$ is an open set of H containing f, disjoint from $S(\lambda, \mu)$. Thus $S(\lambda, \mu)$ is a nonvoid closed set. It follows from (iii) that the $S(\lambda, \mu)$ have the finite intersection property, and thus (exercise 11.11) there exists $f \in H$ belonging to all $S(\lambda, \mu)$. Then $x_\lambda = f(\lambda)$ defines the required family. ●

To prove Theorem 11 we need some further notation. If J is a $\bigvee$-representation of $a \in L$, we call $|J|$ the *rank* of the representation and $\sum (|X| \mid X \in J)$ the *length* of the representation.

If $H \subseteq L$, then a $\bigvee$-*representation* of H, $J(H)$, is a family $(J_a \mid a \in H)$, where J_a is a $\bigvee$-representation of a. If n is an integer and rank $J_a = n$ for each $a \in H$, then $J(H)$ is said to have rank n. A $\bigvee$-representation $J(H)$ of H is said to be *special* if:

(i) $a \in H$, $X, Y \in J_a$, and $\bigwedge X \le \bigwedge Y$ imply that $X = Y$.

(ii) $a,b \in H$ and $a \le b$ imply that for each $X \in J_a$ there is a $Y \in J_b$ such that $\bigwedge X \le \bigwedge Y$.

Each $H \subseteq L$ has a special $\bigvee$-representation; by Theorem 9, J_a need only be chosen as a normal $\bigvee$-representation of a for each $a \in H$.

PROOF OF THEOREM 11. Let us assume that the conclusion of the theorem fails to hold—that is, that there exists an uncountable chain C in L. Let $J(C)$ be a special $\bigvee$-representation of C and let

$$C_n = \{a \mid a \in C, \quad \text{rank } J_a = n\}.$$

Then $C = \bigcup (C_n \mid 1 \le n < \omega)$, and thus some C_n is uncountable. In other words, we can choose C so that rank $J_a = n$ for all $a \in C$. We prove that this is impossible by induction on n.

Let $n = 1$. For each integer k, let

$$C^{(k)} = \{a \mid a \in C, \text{length } J_a = k\}.$$

For each $a \in C^{(k)}$, let $J(a) = \{X_a\}$ and let

$$I_a = \{i \mid i \in I, X_a \cap L_i \ne \varnothing\}.$$

Since any $a,b \in C^{(k)}$ are comparable and $|I_a| = |I_b| = k$, we conclude by Corollary 7 that $I_a = I_b$. Set $I_k = I_a$ for all $a \in C^{(k)}$. For each $i \in I_k$ the set

$$H_i = \{x \mid x \in X_a \cap L_i, a \in C^{(k)}\}$$

is a chain in L_i and is therefore countable. Since $C^{(k)}$ is isomorphic to a chain in $\prod (H_i \mid i \in I_k)$ and I_k is finite, $C^{(k)}$ itself is countable. Because $C = \bigcup (C^{(k)} \mid k < \omega)$, C is countable.

Now let us assume that if C is a chain in L, and $J(C)$ is a special $\bigvee$-

representation of C such that rank $J_a = k$ for all $a \in C$, $k < n$, then C is countable. Let C be a chain in L and let $J(C)$ be a special $\bigvee$-representation of C such that rank $J_a = n$ for all $a \in C$. For $a,b \in C$, $a \leq b$, we define a relation $\varphi_{ab} \subseteq J_a \times J_b$: For $X \in J_a$, $Y \in J_b$, $\langle X, Y \rangle \in \varphi_{ab}$ iff $\bigwedge X \leq \bigwedge Y$. Then φ_{ab} satisfies (i) of Lemma 12 by the second clause in the definition of special $\bigvee$-representations; (ii) and (iii) of Lemma 12 are trivially satisfied. Thus, by Lemma 12, there is a family $\mathscr{X} = (X_a \mid a \in C, X_a \in J_a, J_a \in J(C))$ such that $\bigwedge X_a \leq \bigwedge X_b$ for $a \leq b$. Then $\mathscr{X}$ is a $\bigvee$-representation of rank 1 of a chain, and therefore $\mathscr{X}$ is countable. Thus there are uncountably many elements in C but only countably many X_a; consequently, there is an uncountable subchain C_0 of C such that $X_a = X_b$ for $a,b \in C_0$. The family $\mathscr{J} = (J_a - \{X_a\} \mid a \in C_0)$ is uncountable and of rank $n - 1$. $\mathscr{J}$ is clearly a $\bigvee$-representation of some uncountable subset $H \subseteq L$, and the validity of clause (i) of the definition of a special representation is clear. Now let $a,b \in C_0$, $a \leq b$, and let $X \in J_a - \{X_a\}$. Then there is a $Y \in \dot{J}_b$ such that $\bigwedge X \leq \bigwedge Y$. If $Y = X_b$, then, since $X_a = X_b$, $\bigwedge X \leq \bigwedge X_a$, contradicting the fact that $J(C)$ is special. Thus $Y \in J_b - \{X_a\}$ and so H is a chain with a special $\bigvee$-representation $\mathscr{J}$. However, rank $\mathscr{J} = n - 1$ and H is uncountable, contradicting the induction hypothesis. ⬤

Exercises

In the first paragraph of this section it was shown how Theorem 5.24 can be used to show the existence of free distributive products. The same method, however, does not apply to distributive lattices with 0 and 1. Nevertheless, the idea of the proofs of Theorems 5.5 and 5.24 can be used to get a much stronger result on the existence of free products; it is easiest to formulate this result (G. Grätzer [1968], Theorem 29.2) for universal algebras. To do so we have to introduce some concepts.

A *type* τ of algebras is a sequence $\langle n_0, n_1, \ldots, n_\gamma, \ldots \rangle$ of non-negative integers, $\gamma < o(\tau)$, where $o(\tau)$ is an ordinal called the *order of* τ. An *algebra* $\mathfrak{A}$ *of type* τ is an ordered pair $\langle A; F \rangle$, where A is a nonvoid set and F is a sequence $\langle f_0, \ldots, f_\gamma, \ldots \rangle$, $\gamma < o(\tau)$, where f_γ is an n_γ-ary operation on A. If $o(\tau)$ is finite, $o(\tau) = n$, then we write $\langle A; f_0, \ldots, f_{n-1} \rangle$ for $\langle A; F \rangle$.

1. Define the concepts of subalgebra, polynomial, identity, and equational class for algebras of a given type τ. Show that if **K** is an equational class, $\mathfrak{A}$ is an algebra in **K**, and $\mathfrak{B}$ is a subalgebra of $\mathfrak{A}$, then $\mathfrak{B}$ is in **K**.

2. Define the concepts of homomorphism, homomorphic image, and direct product for algebras of a given type. Show that an equational class is

closed under the formation of homomorphic images and direct products.

3. Let $\mathfrak{A} = \langle A; F \rangle$ be an algebra, let $H \subseteq A$, and let $H \neq \varnothing$. Show that there exists a smallest subset $[H]$ of A, $[H] \supseteq H$ such that $\langle [H]; F \rangle$ is a subalgebra of $\mathfrak{A}$. (This subalgebra is said to be *generated by H*.)

4. Show that $|[H]| \leq |H| + |F| + \aleph_0$.

5. Modify Definition 1 for algebras. Show that the φ in (ii) is unique.

6. Let $\mathfrak{B}$ and $\mathfrak{C}$ be free K-products of $\mathfrak{A}_i$, $i \in I$, with the embeddings ϵ_i, $i \in I$, and χ_i, $i \in I$, respectively. Show that there exists an isomorphism $\alpha: B \to C$ such that $\epsilon_i \alpha = \chi_i$ and $\chi_i \alpha^{-1} = \epsilon_i$ for all $i \in I$.

7. Let **K** be an equational class of algebras and let $\mathfrak{A}_i \in \mathbf{K}$ for $i \in I$. Choose a set S satisfying

$$|S| \geq \sum |A_i| + |F| + \aleph_0.$$

Let Q be the set of all pairs $\langle \mathfrak{B}, (\varphi_i \mid i \in I) \rangle$ such that $B \subseteq S$, φ_i is a homomorphism of $\mathfrak{A}_i$ into $\mathfrak{B}$, and $B = [\bigcup (A_i\varphi \mid i \in I)]$. Form $\mathfrak{A} = \prod(\mathfrak{B} \mid \langle \mathfrak{B}, (\varphi_i \mid i \in I) \rangle \in Q)$ (direct product), and for $a \in A_i$ define $f_a \in A$ by $f_a(\langle \mathfrak{B}, (\varphi_i \mid i \in I) \rangle) = a\varphi_i$. Finally, let $\mathfrak{N}$ be the subalgebra generated by the f_a, $a \in A_i$, $i \in I$. Show that $\mathfrak{N} \in \mathbf{K}$, $a \to f_a$ is a homomorphism ϵ_i of $\mathfrak{A}_i$ into $\mathfrak{N}$ for $i \in I$ and that $\mathfrak{N}$ is generated by $\bigcup (A_i\epsilon_i \mid i \in I)$.

8. Show that ϵ_i is one-to-one iff, for $i \in I$, $a,b \in A_i$, $a \neq b$, there exists an algebra $\mathfrak{C} \in \mathbf{K}$ and homomorphisms $\psi_j: \mathfrak{A}_j \to \mathfrak{C}$ such that $a\psi_i \neq b\psi_i$.

9. Combine the previous exercises to prove the following result.
   EXISTENCE THEOREM FOR FREE PRODUCTS: *Let* **K** *be an equational class of algebras,* $\mathfrak{A}_i \in \mathbf{K}$ *for* $i \in I$. *A free* K-*product of* $\mathfrak{A}_i$, $i \in I$, *exists iff, for* $i \in I$, $a,b \in A_i$, $a \neq b$, *there exist a* $\mathfrak{C} \in \mathbf{K}$, *homomorphisms* $\psi_i: \mathfrak{A}_j \to \mathfrak{C}$ *for* $j \in I$ *such that* $a\psi_i \neq b\psi_i$.

10. Show that in proving the existence of free distributive products and free {0, 1}-distributive products, we can always choose $\mathfrak{C} = \mathfrak{C}_2$, the two-element chain, in applying exercise 9.

11. Show that the two definitions of free K-product are equivalent for any class **K** of lattices.

12. Show that the free Boolean algebra on $\mathfrak{m}$ generators is a free {0, 1}-distributive product of $\mathfrak{m}$ copies of the free Boolean algebra on one generator.

13. Prove that the free Boolean algebra on $\mathfrak{m}$ generators can be represented by the clopen subsets of $\{0, 1\}^{\mathfrak{m}}$ where $\{0, 1\}$ is the two-element discrete topological space.

14. Find a topological representation for the free distributive lattice on $\mathfrak{m}$ generators (G. Ja. Areškin [1953b]).

The remaining exercises in this Section are based on G. Grätzer and H. Lakser [1969a].

15. Let L_0 and L_1 be bounded distributive lattices and let $a_0,b_0 \in L_0$, $a_1,b_1 \in L_1$, and $a_0 \parallel b_0$, $a_1 \parallel b_1$. Let L be a free {0, 1}-distributive product of L_0 and L_1. Define $x \in L$ by

$$x = (a_0 \wedge a_1) \vee (b_0 \wedge b_1).$$

Find the normal $\bigvee$-representation and also the normal $\bigwedge$-representation of x.

16. Let L_i, $i \in I$, be distributive lattices. For each $i \in I$, let $\bar{L}_i$ be the result of adjoining 0 and 1 to L_i. Let $\bar{L}$ be a free $\{0, 1\}$-distributive product of the $\bar{L}_i$, $i \in I$. Prove that $L = \bar{L} - \{0, 1\}$ is a distributive lattice and that it is a free distributive product of the L_i, $i \in I$.

17. A lattice L satisfies the $\mathfrak{m}$-*chain condition* if all chains in L have cardinality $< \mathfrak{m}$. For an infinite cardinal $\mathfrak{m}$, and a cardinal $\mathfrak{n} > 0$, let us say that the condition $P(\mathfrak{m}, \mathfrak{n})$ is satisfied if, whenever L_i, $i \in I$, are distributive lattices and the L_i satisfy the $\mathfrak{m}$-chain condition, $|I| = \mathfrak{n}$, then a free distributive product L of the L_i, $i \in I$, satisfies the $\mathfrak{m}$-chain condition. Let $P_{\{0, 1\}}(\mathfrak{m}, \mathfrak{n})$ denote the same condition for bounded distributive lattices and the free $\{0, 1\}$-distributive product and let $P_{\mathbf{B}}(\mathfrak{m}, \mathfrak{n})$ be the same condition for Boolean algebras. Observe that Theorem 11 states that $P_{\{0, 1\}}(\aleph_1, \mathfrak{n})$ holds.

   Generalize the proof of Theorem 11 to show that $P_{\{0, 1\}}(\mathfrak{m}, \mathfrak{n})$ holds for any regular cardinal $\mathfrak{m} > \aleph_0$. ($\mathfrak{m}$ is *regular* if $\mathfrak{m}_i < \mathfrak{m}$ for $i \in I$ and $|I| < \mathfrak{m}$ imply that $\Sigma\,(\mathfrak{m}_i \mid i \in I) < \mathfrak{m}$.)

18. Let $\mathfrak{m}$ be an infinite cardinal not cofinal with ω (that is, $\mathfrak{m} = \sum\,(\mathfrak{m}_i \mid i \in I)$ and $|I| = \aleph_0$ imply that $\mathfrak{m} = \mathfrak{m}_i$ for some $i \in I$). Prove that there exists a Boolean lattice B such that B satisfies the $\mathfrak{m}$-chain condition but for all $\mathfrak{n} < \mathfrak{m}$ there exists in B a chain of cardinality $\mathfrak{n}$.

19. Let $\mathfrak{m}$ be *singular* (that is, not regular) and not cofinal with ω. Prove that there exist Boolean lattices B_0 and B_1 satisfying the $\mathfrak{m}$-chain condition such that a free $\{0, 1\}$-distributive product of B_0 and B_1 has a chain of cardinality $\mathfrak{m}$.

20. Let $\mathfrak{m}$ be cofinal with ω. Show that $P_{\mathbf{B}}(\mathfrak{m}, \aleph_0)$ does not hold.

*21. Let $\mathfrak{m}$ be cofinal with ω and let L be a distributive lattice that has a chain of cardinality $\mathfrak{m}_1$ for all $\mathfrak{m}_1 < \mathfrak{m}$. Verify that L has a chain of cardinality $\mathfrak{m}$ (A. Hajnal).

22. Prove that $P(\mathfrak{m}, \mathfrak{n})$ holds iff either $\mathfrak{n} \geq \aleph_0$, $\mathfrak{m}$ is regular, and $\mathfrak{m} > \aleph_0$, or $1 < \mathfrak{n} < \aleph_0$, and $\mathfrak{m}$ is either regular or cofinal with ω.

23. Prove exercise 22 for $P_{\{0, 1\}}(\mathfrak{m}, \mathfrak{n})$.

24. Prove exercise 22 for $P_{\mathbf{B}}(\mathfrak{m}, \mathfrak{n})$.

25. Establish Theorem 11 without using the Axiom of Choice.

13. Some Categorical Concepts

The *category of distributive lattices* **D** is the class of all distributive lattices together with the class of all lattice homomorphisms amongst them. We use the word "category" to emphasize that not only the elements (called *objects*) of **D**, but also the *maps* (called *morphisms*) that belong to the category, are specified. To illustrate this idea, consider the category **B**

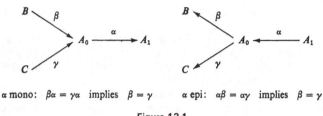

α mono: $\beta\alpha = \gamma\alpha$ implies $\beta = \gamma$ α epi: $\alpha\beta = \alpha\gamma$ implies $\beta = \gamma$

Figure 13.1

of Boolean algebras and the Boolean algebra homomorphisms. A related category is $\mathbf{B}^L$, the elements of which are Boolean lattices and the maps of which are lattice homomorphisms. We will also refer to the category $\mathbf{D}_{\{0,1\}}$ of bounded distributive lattices and $\{0, 1\}$-homomorphisms. For instance, the object $\mathfrak{C}_2$ belongs to $\mathbf{D}$ and $\mathbf{D}_{\{0,1\}}$; in $\mathbf{D}$ there are three maps from $\mathfrak{C}_2$ to $\mathfrak{C}_2$; in $\mathbf{D}_{\{0,1\}}$ there is only one.

Category theory is the study of those properties of algebras (and other objects) that can be expressed in terms of maps. In this section we shall discuss a few such concepts and investigate them in the categories $\mathbf{D}$, $\mathbf{D}_{\{0,1\}}$, and $\mathbf{B}$.

For our purposes, in the subsequent definitions the clauses "let $\mathbf{K}$ be a category" and "let $\mathbf{K}$ be a category of algebras" shall mean: "Let $\mathbf{K}$ be one of $\mathbf{D}$, $\mathbf{D}_{\{0,1\}}$, $\mathbf{B}$." "An object of K" shall mean "a distributive lattice," "a distributive lattice with 0 and 1," and "a Boolean algebra," respectively. "A map" shall mean "a homomorphism" and "a $\{0, 1\}$-homomorphism," respectively. Those who know the definition of a general category or of a category of algebras will realize that the easy observations of this section, but none of the more profound ones, apply to the general case. Some categories of pseudocomplemented lattices will be considered in Chapter 3.

In a general category, the morphisms need not be maps, and therefore such concepts as one-to-one maps and onto maps are not meaningful. Definitions 1 and 2 are obtained by abstracting certain properties of one-to-one and onto maps, respectively.

DEFINITION 1.   *A map* $\alpha: A_0 \to A_1$ *in the category* $\mathbf{K}$ *is called* mono, *or a* monomorphism, *iff, for objects B, C of* $\mathbf{K}$ *and maps* $\beta: B \to A_0$, $\gamma: C \to A_0$, *if* $\beta\alpha = \gamma\alpha$, *then* $\beta = \gamma$ *(and B = C)* (see Figure 13.1).

DEFINITION 2.   *A map* $\alpha: A_1 \to A_0$ *in the category* $\mathbf{K}$ *is called* epi, *or an* epimorphism, *iff, for objects B, C of* **K** *and maps* $\beta: A_0 \to B, \gamma: A_0 \to C$, *if* $\alpha\beta = \alpha\gamma$, *then* $\beta = \gamma$ (*and B = C*) (see Figure 13.1).

LEMMA 3. *Let* **K** *be a category of algebras, and let us assume the existence of the free algebra F over* **K** *with one free generator* x_0. *Then monomorphisms in* **K** *are exactly the one-to-one maps in* **K**.

PROOF.   The reader can easily verify that a one-to-one map is a monomorphism. Now let $\alpha: A_0 \to A_1$ be not one-to-one—that is, let $a\alpha = b\alpha$ for some $a \neq b$, $a,b \in A_0$. Let β and γ be the homomorphisms of F into A_0, extending the maps $x_0 \to a$ and $x_0 \to b$, respectively. Then $\beta\alpha = \gamma\alpha$ but $\beta \neq \gamma$, so α is not a monomorphism.   ●

Since each category of **D**, $\mathbf{D}_{\{0,1\}}$, and **B** satisfies the condition of Lemma 3, monomorphisms are one-to-one homomorphisms in each of these categories.

Epimorphisms, however, behave in a very interesting way. Let $A_0, A_1 \in \mathbf{D}$ and $\alpha: A_0 \to A_1$ be given as in Figure 13.2. Obviously, α is a homomorphism and α is not onto. Now let $A_2 \in \mathbf{D}$, $\beta: A_1 \to A_2$, and $\gamma: A_1 \to A_2$ such that $\alpha\beta = \alpha\gamma$. Then $b\beta = a(\alpha\beta) = a(\alpha\gamma) = b\gamma$; similarly, $0\beta = 0\gamma$ and $1\beta = 1\gamma$. Therefore, $c\beta$ and $c\gamma$ both satisfy in A_2 the equations $b\beta \wedge x = 0\beta$ and $b\beta \vee x = 1\beta$; thus, by Corollary 7.3, $c\beta = c\gamma$. This verifies that $\beta = \gamma$—in other words, that α is epi. The next result shows that this example is typical.

For a lattice L and $H \subseteq L$, let $[H]_B$ denote the smallest sublattice of L closed under relative complementation in L. Note that if L is a generalized Boolean lattice, then $[H \cup \{0\}]_B = [H]_R$.

THEOREM 4.   *In the categories* **D** *and* $\mathbf{D}_{\{0,1\}}$, *the map* $\alpha: A_0 \to A_1$ *is epi iff* $[A_0\alpha]_B = A_1$.

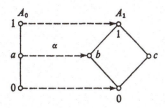

Figure 13.2

PROOF. Let $\beta_0: A_1 \to C_0, \beta_1: A_1 \to C_1$. Then $\alpha\beta_0 = \alpha\beta_1$ means that β_0 and β_1 agree on $A_0\alpha$; but then they agree on $A_2 = [A_0\alpha]_B$. Therefore, if $A_2 = A_1$, then $\beta_0 = \beta_1$, and α is epi. Conversely, if $A_2 \neq A_1$, then adjoin 0, 1 (if there were none) and form the Boolean algebras B_1 and B_2 R-generated by A_1 and A_2, respectively. $A_1 \neq A_2$ implies that $B_1 \neq B_2$ by Lemma 10.9. Take an $a \in B_1 - B_2$. By Lemma 3.5 and Theorem 7.15, we can take a homomorphism β of B_2 into $\mathfrak{C}_2$ such that for $x < a, x\beta = 0$, and for $x > a, x\beta = 1$. By Theorem 8.5, β can be extended to $\beta_i: [B_2 \cup \{a\}]_B \to \mathfrak{C}_2$ such that $a\beta_i = i, i = 0, 1$. If we can extend the β_i to $\beta_i: B_1 \to \mathfrak{C}_2$, then $\alpha\beta_0 = \alpha\beta_1, \beta_0 \neq \beta_1$, showing that α is not epi. This will be accomplished by the following lemma:

LEMMA 5. *Let B_1, B_2, B be Boolean algebras, let B_2 be a subalgebra of B_1, and let B be complete. Let $\varphi: B_2 \to B$ be a homomorphism. Then φ can be extended to a homomorphism $\psi: B_1 \to B$.*

PROOF. Let us form the partially ordered set Q whose elements are ordered pairs $\langle C, \gamma \rangle$, where C is a subalgebra of B_1, $C \supseteq B_2$, γ is a homomorphism, $\gamma: C \to B$, and γ extends φ. Define $\langle C_0, \gamma_0 \rangle \leq \langle C_1, \gamma_1 \rangle$ to hold iff C_0 is a subalgebra of C_1 and γ_1 extends γ_0. Zorn's Lemma can be applied to Q, thus yielding a maximal element $\langle M, \psi \rangle$. Since Corollary 8.7 would contradict $M \neq B_1$, we must have $M = B_1$. ●

Theorem 4 yields

COROLLARY 6. *In the category $\mathbf{B}$ the epimorphisms are exactly the onto maps in $\mathbf{B}$.*

Besides selecting interesting maps, category theory also singles out interesting objects.

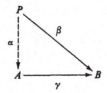

Figure 13.3

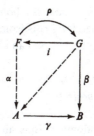

Figure 13.4

DEFINITION 7. *Let* **K** *be a category of algebras. An object P of* **K** *is called projective if, for any onto map* $\gamma: A \to B$, *and any map* $\beta: P \to B$, *there is a map* $\alpha: P \to A$ *with* $\alpha\gamma = \beta$ (*see Figure 13.3*).

Projective algebras generalize a property of free algebras. Note that the categorical definition of projectivity requires γ to be epi; this definition is not interesting in **D** (see exercise 4).

LEMMA 8. *Let F be a free algebra in* **K**. *Then F is projective.*

PROOF. Let x_i, $i \in I$, be the free generators of F. Let $a_i \in (x_i\beta)\gamma^{-1}$ and let α be the homomorphism extending $x_i \to a_i$, $i \in I$. ●
  Life would be much simpler if the converse of Lemma 6 were also true. However, it does not hold in any of our categories.
  Let us call a homomorphism ρ of A into A an *endomorphism*; let us call an endomorphism ρ a *retraction* if $\rho^2 = \rho$—in other words, if ρ is the identity map on $A\rho$. $A\rho$ is then called a *retract* of A.

LEMMA 9. *Let F be a free algebra in* **K** *and let G be a retract of F. Then G is projective.*

PROOF. The situation is shown in Figure 13.4. ρ is the retraction, i is the inclusion map of G into F, γ is a map of A onto B. Since $\rho\beta: F \to B$ and F is projective, there is a map $\alpha: F \to A$ with $\alpha\gamma = \rho\beta$. Therefore, $(i\alpha)\gamma = i(\alpha\gamma) = i(\rho\beta) = (i\rho)\beta = \beta$, since $i\rho$ is the identity map on G. Thus $(i\alpha)\gamma = \beta$, and G is projective. ●
  The converse is equally trivial. If G is projective, then if we take a free algebra F with $|G|$ free generators x_i, $i \in I$, we can find an onto map $\{x_i \mid i \in I\} \to G$ that extends to a homomorphism γ. Let β be the identity

map on G. Since G is projective, there is a homomorphism $\alpha\colon G \to F$, such that $\alpha\gamma = $ identity on G. Therefore, G is isomorphic to a retract $G\alpha$ of F, and by stretching this concept a bit, we can say that G is a retract of F.

LEMMA 10. *If free algebras exist on any set of generators, then only retracts of free algebras are projective.*

It is easy to construct endomorphisms and retractions in distributive lattices; thus Lemma 10 is only the beginning of the real work in determining the projectives.

THEOREM 11 (R. Balbes [1967]). *A finite distributive lattice P is projective (in* D) *iff the join of any two meet-irreducible elements is meet irreducible.*

PROOF.[8] Let P be projective. By Lemma 10 we can assume that $P \subseteq F$, where F is free on x_i, $i \in I$, and ρ is a retraction: $F \to P$. Let a and b be meet-irreducible and let $a \vee b$ be meet-reducible—that is, for some $c,d \in P$, $a \vee b \geq c \wedge d$, $a \vee b \nleq c$, $a \vee b \nleq d$. Let $c = \bigvee (\bigwedge C_k \mid k \in K)$ and let $d = \bigvee (\bigwedge D_l \mid l \in L)$, where C_k and D_l are finite subsets of $\{x_i \mid i \in I\}$. Then $c = c\rho = \bigvee (\bigwedge C_k\rho \mid k \in K)$, $d = d\rho = \bigvee (\bigwedge D_l\rho \mid l \in L)$. Therefore, k and l exist such that $\bigwedge C_k\rho \nleq a \vee b$ and $\bigwedge D_l\rho \nleq a \vee b$. At the same time, $\bigwedge C_k \wedge \bigwedge D_l \leq c \wedge d \leq a \vee b$; therefore, $\bigwedge C_k \wedge \bigwedge D_l \leq a$ or $\bigwedge C_k \wedge \bigwedge D_l \leq b$ by Theorem 8.3. Applying ρ we get $\bigwedge C_k\rho \wedge \bigwedge D_l\rho \leq a$ or $\bigwedge C_k\rho \wedge \bigwedge D_l\rho \leq b$, which means that either a or b is meet-reducible, a contradiction.

Conversely, let P satisfy the condition of the theorem and let m_i, $i \in I$, be the meet-irreducible elements of P. Let F be the free distributive lattice on x_i, $i \in I$, and let α be the homomorphism of F onto P extending $x_i \to m_i$, $i \in I$. Let G be the join-subsemilattice of F generated by the x_i, $i \in I$. Now we define a map $\varphi\colon P \to F$ by

$$a\varphi = \bigwedge (x \mid x \in G, x\alpha \geq a) \quad \text{for } a \in P.$$

In other words, $a\varphi$ is the smallest element in the subset $a\alpha^{-1}$ of F.

Distributivity shows that $(a \vee b)\varphi = a\varphi \vee b\varphi$, since any x in G such that $x\alpha \geq a \vee b$ is of the form $x = y \vee z$, $y \in G$, $y\alpha \geq a$, $z \in G$, $z\alpha \geq b$, and conversely. To show that φ preserves meets, note that, by the assump-

---

[8] This is a simplified version of R. Balbes [1967] taken from G. Grätzer and B. Wolk [1970]. However, the crucial map φ is, by necessity, the same as that of Balbes.

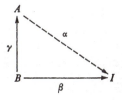

Figure 13.5

tions of the theorem, $x\alpha$ is meet-irreducible for all x in G. Thus $x\alpha \geq a \wedge b$ is equivalent to the condition $x\alpha \geq a$ or $x\alpha \geq b$. This proves that φ is a homomorphism of P into F.

Finally, we have $a = \bigwedge (m_i \mid m_i \geq a)$ for every element a of P. Consequently, $a\varphi\alpha = a$; therefore, P is a retract of F and is thus projective by Lemma 9. ⬤

Not much is known about projectives in the category **B**.

LEMMA 12. *Every countable Boolean algebra is projective.*

PROOF.   Let B be a countable Boolean algebra. Then by Theorem 10.20, $B = B(C)$, where C is a countable chain with 0 and 1. Let $C - \{0, 1\} = \{c_0, c_1, \ldots, c_n, \ldots\}$ and let F be the free Boolean algebra on $\aleph_0$ generators, $x_0, x_1, \ldots, x_n, \ldots$. We define a map $\rho: C \to F$. Let $c_0\rho = x_0$, $0\rho = 0$, and $1\rho = 1$. Suppose that $c_0\rho, \ldots, c_{n-1}\rho$ have already been defined and choose $d,e \in \{0, 1, c_0, \ldots, c_{n-1}\}$ such that in the sublattice $\{0, 1, c_0, \ldots, c_n\}$, $d \prec c_n \prec e$. Set $c_n\rho = (d\rho \vee x_n) \wedge e\rho$. Let $\bar{\rho}$ be the extension of ρ to a homomorphism: $B \to F$. Let φ be the homomorphism of F onto B satisfying $x_i\varphi = c_i$. Then $\bar{\rho}\varphi$ is the identity on B, and thus B is a retract of F. ⬤

Reversing the arrows in Definition 7 and replacing "onto" by "one-to-one," we get another interesting concept:

DEFINITION 13. *Let **K** be a category of algebras. The object I is called injective if, for any one-to-one map $\gamma: B \to A$ and map $\beta: B \to I$, there is a map $\alpha: A \to I$ with $\gamma\alpha = \beta$ (see Figure 13.5).*

Intuitively, a homomorphism into I from a subalgebra can be extended to the whole algebra. Injectives were first studied for modules; however, they turn out to be very interesting in our categories. In fact, injective objects in all three categories are the same:

THEOREM 14. *In the categories* **D**, **D**$_{(0, 1)}$, *and* **B**, *the injective objects are the same: They are the complete Boolean lattices (algebras).*

The proof of this theorem is based on a categorical triviality and a simple lattice theoretic lemma.

LEMMA 15. *Let I be injective. Then I is a retract of every extension of I.*

PROOF. Let I be a subalgebra of A, let $B = I$, and let β and γ be the identity map of I. Then there exists an $\alpha: A \to I$ such that $\gamma\alpha = \beta$, that is, for $x \in I$, $x\alpha = x$. In other words, α is a retraction. ●

LEMMA 16. *Let A be a retract of a complete Boolean lattice B. Then A is a complete Boolean lattice.*

PROOF. Let α be the retraction, $u = 1\alpha$, and $v = 0\alpha$. Then $0 \le x \le 1$ for all $x \in A$, and therefore $v \le x\alpha = x \le u$; thus u is the largest and v is the smallest element of A. Now let $x \in A$, $x \wedge y = 0$, and $x \vee y = 1$ ($y \in B$). Then $x \wedge y\alpha = v$, $x \vee y\alpha = u$, and thus A is a Boolean lattice.

Now let $X \subseteq A$ and $a = \bigvee X$ in B. Then $x \le a$ for all $x \in X$; thus $x\alpha = x \le a\alpha$. If, for $y \in A$, $x \le y$ for all $x \in X$, then $a \le y$, and therefore $a\alpha \le y$, showing that $\bigvee X = a\alpha$ in A. Thus A is complete. ●

PROOF OF THEOREM 14. Let I be injective in one of **D**, **D**$_{(0, 1)}$, **B**. Then I is a distributive lattice, and thus (by Theorem 10.1) it has an extension B, which is a complete (atomic) Boolean lattice. By Lemma 15, I is a retract of B; therefore, by Lemma 16, I is a complete Boolean lattice.

Conversely, if I is a complete Boolean lattice, then I is injective in **B** by Lemma 5. Since (Corollary 10.7) every homomorphism $\varphi: L_0 \to L_1$ of bounded distributive lattices can be extended to a homomorphism of the Boolean lattices they generate, I is injective in **D** and **D**$_{(0, 1)}$ also. ●

In the category **S** of sets and maps,
 (i) every epimorphism has a left inverse that is mono, and
 (ii) every monomorphism has a right inverse that is epi.
G. Birkhoff [1967] asks whether (i) or (ii) holds for **B** or **D**$_{fin}$ (the category of finite distributive lattices and lattice homomorphisms).

Using the results of this section we can easily answer this question. By Corollary 6, (i) means that in **B** every homomorphic image can be regarded as a retract, which is obviously not the case since a complete Boolean

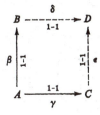

Figure 13.6

algebra can have noncomplete homomorphic images (see exercise 17); it can have only complete retracts (Lemma 16). Similarly, (ii) means that every subalgebra is a retract, which is false for the same reason. Thus in **B** neither (i) nor (ii) holds.

In $\mathbf{D}_{\mathrm{fin}}$ let $\alpha\colon \mathfrak{C}_3 \to (\mathfrak{C}_2)^2$ be one-to-one. Therefore, α is epi by Theorem 4. If β is a left inverse of α and β is mono, then β is one-to-one from a four-element set into a three-element set, a contradiction. Thus (i) fails in $\mathbf{D}_{\mathrm{fin}}$ and so does (ii): Any distributive lattice is a sublattice of a Boolean lattice, but only the Boolean ones are retracts. Thus in $\mathbf{D}_{\mathrm{fin}}$ also, neither (i) nor (ii) holds.

Finally, we consider a categorical property not of maps or algebras but of whole classes of algebras.

DEFINITION 17 (B. Jónsson [1961] and [1965]). *Let* **K** *be a class of algebras.* **K** *is said to have the* Amalgamation Property *if, whenever $A,B,C \in \mathbf{K}$ with A a subalgebra of B and of C, then there exists an algebra $D \in \mathbf{K}$ containing up to isomorphism B and C as subalgebras such that $B \cap C \supseteq A$.*

An equivalent definition is illustrated by Figure 13.6. If $\beta\colon A \to B$ and $\gamma\colon A \to C$ are one-to-one, then there exist D and one-to-one maps $\delta\colon B \to D$ and $\epsilon\colon C \to D$ in **K** such that $\beta\delta = \gamma\epsilon$.

Many important classes of algebras have this property and many others do not. A simple condition equivalent to the Amalgamation Property is given in the following result.

THEOREM 18. *Let* **K** *be an equational class of algebras.* **K** *has the Amalgamation Property if, whenever $A,B,C \in \mathbf{K}$ with A a subalgebra of B and of C, $a,b \in B$, $a \neq b$, then there exist an algebra $D \in \mathbf{K}$ and homomorphisms $\varphi\colon B \to D$, $\psi\colon C \to D$ such that for every $t \in A$, $t\varphi = t\psi$ and $a\varphi \neq b\varphi$ (see Figure 13.7).*

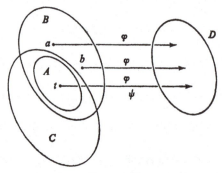

Figure 13.7

PROOF. The "only if" part is obvious. To prove the "if" part, again, all we have to do is to copy the proof of Theorem 5.5; the set S has to be of cardinality $|B| + |C| + |F| + \aleph_0$, and we shall take all pairs $\langle G, \{\beta, \gamma\} \rangle$ where $G \in \mathbf{K}$, $G \subseteq S$, $\beta \colon B \to G$, $\gamma \colon C \to G$ are homomorphisms agreeing on A and where G is generated by $B\beta \cup C\gamma$. The details are left to the reader. ●

COROLLARY 19 (R. S. Pierce [1968]). *Let $\mathbf{K}$ be an equational class of algebras in which every algebra can be embedded in an injective algebra. Then $\mathbf{K}$ has the Amalgamation Property.*

PROOF. Let A, B, and C be given as in Figure 13.7, $a, b \in B$, and $a \neq b$. Let φ be an embedding of B in an injective algebra D; then $a\varphi \neq b\varphi$. The restriction φ_1 of φ to A can be extended to a homomorphism ψ of C into D, since D is injective. Thus the condition of Theorem 18 is satisfied. ●

By Theorem 14, Theorem 7.19, and Corollary 7.21, the condition of Corollary 19 holds in $\mathbf{D}$, $\mathbf{D}_{\{0, 1\}}$, and $\mathbf{B}$. Therefore,

COROLLARY 20. $\mathbf{D}$, $\mathbf{D}_{\{0, 1\}}$, *and* $\mathbf{B}$ *satisfy the Amalgamation Property.*

Compare this result with exercises 14 and 16.

To conclude this section, we give an interesting application of the Amalgamation Property.

THEOREM 21 (B. Jónsson [1961] and [1965]). *Let $\mathbf{K}$ be a category of algebras satisfying the Amalgamation Property, $A, B, C \in \mathbf{K}$ with C a free $\mathbf{K}$-*

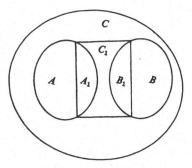

Figure 13.8

*product of A and B. (We assume that A and B are subalgebras of C.) Let A_1
be a subalgebra of A, let B_1 be a subalgebra of B, and let C_1 be the subalgebra
of C generated by A_1 and B_1. If a free $\mathbf{K}$-product of A_1 and B_1 exists, then C_1
is a free $\mathbf{K}$-product of A_1 and B_1 (see Figure 13.8).*

PROOF.   Let C_2 be a free $\mathbf{K}$-product of A_1 and B_1. Thus there exists a
homomorphism χ of C_2 into C_1 that is the identity on A_1 and B_1. Since A_1
is a subalgebra of A and of C_2, by the Amalgamation Property there is an
algebra D containing A and C_2 as subalgebras, $A_1 \subseteq A \cap C_2$. Similarly,
B_1 is a subalgebra of D and B, and thus there exists an algebra E containing
B and D as subalgebras, $B_1 \subseteq B \cap D$ (see Figure 13.9). Since C is a free
product of A and B, there exists a homomorphism φ_1 of C into E that is
the identity on A and B. Let φ be the restriction of φ_1 to C_1. Then φ maps
C_1 into C_2, $\varphi\chi$ is the identity on A_1 and B_1, and $\varphi\chi$ is thus the identity on
C_1. Similarly, $\chi\varphi$ is the identity on C_2, so φ is an isomorphism between C_1
and C_2.   ●

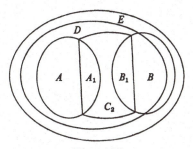

Figure 13.9

Exercises

1. Let **K** be the category of one- and two-element Boolean algebras, with all homomorphisms as morphisms. Which maps are epi and which are mono?

2. Prove that if α and β are epi (mono), then so is $\alpha\beta$.

3. An isomorphism is both epi and mono. Is the converse also true?

4. In Definition 7, change "onto" to "epi." Show that in **D** no lattice satisfies this definition.

5. Show that a direct product of injective algebras is injective.

6. Show that a free product of projective algebras is projective.

7. Is a retract of an injective algebra injective?

8. Is a retract of a projective algebra projective?

9. Let **K** be a category of algebras, $A, A_0, A_1 \in \mathbf{K}$. Let the algebra A be the direct product of A_0 and A_1. Assume that for any $X, Y \in \mathbf{K}$ there is at least one homomorphism $\varphi: X \to Y$. Show that A is injective iff A_0 and A_1 are injective.

10. Let A, B, and C be lattices, let A be a sublattice of B and of C, and let $B \cap C = A$. Define a partial order on $B \cup C$ as follows: For $x, y \in B$ (for $x, y \in C$), $x \leq y$ iff $x \leq y$ in B (in C); for $x \in B$, $y \in C$ (and for $x \in C$, $y \in B$), $x \leq y$ iff for some $z \in A$, $x \leq z$ in B (in C) and $z \leq y$ in C (in B). Show that $B \cup C$ is a partial lattice.

11. Prove that the Amalgamation Property holds for the category **L** of all lattices.

12. Let **K** be an equational class of algebras. The *Strong Amalgamation Property* holds in **K** if, whenever $A, B, C \in \mathbf{K}$, and A is a subalgebra of B and of C, there exists an algebra D in **K** containing B and C as subalgebras (up to isomorphism) such that $B \cap C = A$. Show that the Strong Amalgamation Property holds in **K** iff, in addition to the condition of Theorem 18, the following condition holds:
If $a \in B - A$ and $b \in C - A$, then there exists an algebra $D \in \mathbf{K}$ and homomorphisms $\varphi: B \to D$ $\psi: C \to D$ such that for all $t \in A$, $t\varphi = t\psi$ and $a\varphi \neq b\psi$.

13. Show that the Strong Amalgamation Property holds for the class of all lattices.

14. Show that the Strong Amalgamation Property holds for **B**.

15. Let A, B, and C be bounded distributive lattices, let A be a $\{0, 1\}$-sublattice of B and of C, and let $B \cap C = A$. Show that there exists a bounded distributive lattice D containing B and C as $\{0, 1\}$-sublattices such that $B \cap C = A$ iff A is closed under relative complementation in B and C.

16. Prove that the Strong Amalgamation Property fails in $\mathbf{D}_{\{0, 1\}}$ and **D**.

17. Let B be the Boolean algebra of all subsets of an infinite set I. Show that, although B is complete, B/J is not complete if J is the ideal of all finite subsets of I.

18. Show that Theorem 18 holds for any class closed under direct products.

19. Let **K** be a category of algebras. Show that if the Strong Amalgamation Property holds for **K**, then every epi is onto.
20. Is the converse of exercise 19 true?

Further Topics and References

It seems very hard to generalize the theorem on the uniqueness of an irredundant representation of an element of a finite distributive lattice (Corollary 7.13). The best generalization in this direction is that of R. P. Dilworth and P. Crawley [1960] to distributive algebraic lattices in which every interval has an atom. See the survey article by R. P. Dilworth [1961]; see also S. Kinugawa and J. Hashimoto [1966].

For special classes of lattices, Theorem 7.1 has various stronger forms that claim the existence of copies of $\mathfrak{N}_5$ and/or $\mathfrak{M}_5$ that are very large or very small. For instance, in a bounded relatively complemented non-modular lattice, $\mathfrak{N}_5$ can be chosen so that the o and i of $\mathfrak{N}_5$ are the 0 and 1 of the lattice. The same is true of $\mathfrak{M}_5$ in certain complemented modular lattices (such results follow from J. von Neumann [1936]). If the lattice is finite and modular, then $\mathfrak{M}_5$ can be chosen so that a, b, and c cover o and i covers a, b, and c. If L is finite and nonmodular, then $\mathfrak{N}_5$ can be minimized by requiring that $a \succ b$.

Using the terminology of Section 16, Theorem 7.19 states that every distributive lattice is a subdirect product of copies of $\mathfrak{C}_2$. Distributive lattices that are subdirect products of copies of $\mathfrak{C}_n$ are described in F. W. Anderson and R. L. Blair [1961].

A more general form of Theorem 8.5 can be found in R. Sikorski [1964]; all such theorems can be derived from a universal algebraic triviality; see, for example, Theorem 12.2 in G. Grätzer [1968].

Let L be a lattice and let f be an n-ary function on L, that is, $f: L^n \to L$. We say that f has the *Congruence Substitution Property* if, for every congruence relation Θ in L and $a_i, b_i \in L$, $1 \le i \le n$, $a_i \equiv b_i(\Theta)$, $1 \le i \le n$, imply that $f(a_1, \ldots, a_n) \equiv f(b_1, \ldots, b_n)(\Theta)$. On a Boolean algebra B, a function has the Congruence Substitution Property iff it is a Boolean algebraic function, that is, a Boolean polynomial in which elements of B are substituted for some variables. Functions satisfying the Congruence Substitution Property on a bounded distributive lattice were described in G. Grätzer [1964].

Many properties of ideals of a distributive lattice can be generalized to certain ideals of a general lattice. One such concept is that of *standard ideal* (G. Grätzer [1959], G. Grätzer and E. T. Schmidt [1961]). The method given in Theorem 9.9 is not the only one used to introduce ring operations in a generalized Boolean lattice. G. Grätzer and E. T. Schmidt [1958e] proved that ring operations $+$, $\cdot$ can be introduced on a distributive lattice L such that $+$ and $\cdot$ satisfy the Congruence Substitution Property iff L is relatively complemented. Furthermore, $+$ and are uniquely determined by the zero of the ring, which can be an arbitrary element of L.

Whether every distributive algebraic lattice is isomorphic to the congruence lattice of some lattice is one of the longest-standing problems of lattice theory. The method used in Section 9 for the finite case can be easily extended to infinite algebraic lattices in which every element is a finite join of join-irreducible elements. A further extension of this result is that of E. T. Schmidt [1962] and [1968]. (In reading the two papers, the reader should disregard the Theorem and Lemmas 9 and 10 of the first paper.)

Congruence lattices of distributive lattices were not considered in this text because the problem is trivial by Lemma 10.5: A lattice is isomorphic to the congruence lattice of a distributive lattice iff it is isomorphic to $I(B)$, where B is a generalized Boolean lattice; $I(B)$, by Theorem 9.13, is characterized as a distributive algebraic lattice in which the compact elements form a relatively complemented sublattice.

Algebraic lattices originated in A. Komatu [1943], L. Nachbin [1949], and J. R. Büchi [1952]. The original definition was presented as follows: (i) L is complete; (ii) in L every element is the join of join-inaccessible elements; (iii) L is join-continuous. An element a of L is *join-accessible* if there is a nonvoid subset H of L that is *directed* (for $x,y \in H$ there exists an upper bound $z \in H$), if $\bigvee H = a$, and if $a \notin H$; otherwise, a is *join-inaccessible*. L is *join-continuous* if, for any $a \in L$ and directed $H \subseteq L$, we have $a \wedge \bigvee H = \bigvee (a \wedge h \mid h \in H)$.

Interestingly, it is sufficient to formulate conditions (i) and (iii) of the previous paragraph for chains only. In other words, a lattice L is complete iff $\bigwedge C$ and $\bigvee C$ exist for any chain C of L; and a (complete) lattice L is join-continuous iff $a \wedge \bigvee C = \bigvee (a \wedge c \mid c \in C)$ for any chain C of L. These statements are immediate consequences of the following result of T. Iwamura [1944]. Let H be an infinite directed set. Then H has a decomposition $H = \bigcup (H_\gamma \mid \gamma < \alpha)$, where each H_γ is directed; for $\gamma < \delta < \alpha$ we have $H_\gamma \subseteq H_\delta$, and for $\gamma < \alpha$ we have $|H_\gamma| < H$. Extensions of lattices preserving (iii) were considered in D. A. Kappos and F. Papangelou [1966].

The problem of completions of a lattice has been extensively studied. The standard method is the MacNeille completion (see exercises 10.9–10.14, which are from H. M. MacNeille [1937]). This method is described in detail in G. Bruns [1962a]; see also Y. Sampei [1953]. One of the important shortcomings of this method is that it does not preserve identities, nor even distributivity; see M. Cotlar [1944] and N. Funayama [1944]. What is preserved for complemented modular lattices is examined in J. E. McLaughlin [1961].

If we define a *completion* $\hat{L}$ of a lattice L as any lattice $\hat{L}$ containing L as a sublattice such that all infinite meets and joins that exist in L are preserved in $\hat{L}$, we can ask whether there is *any* distributive completion of a distributive lattice. The answer, in general, is in the negative; see P. Crawley [1962].

For Boolean algebras, set representations preserving all joins and meets, or joins and meets of certain types, play an important role, especially in relationship with infinite distributive identities. The larger part of the book by R. Sikorski [1964] is devoted to this problem. For distributive lattices the questions become even more complicated—see, for instance, G. Bruns [1959], [1961], and [1962b]; C. C. Chang and A. Horn [1962]; A. Horn [1962]; and G. N. Raney [1952], [1953].

Cardinalities of maximal chains in Boolean algebras were considered by J. Jakubik [1957] and [1958]; see also exercise 10.28.

Utilizing the sequence I_γ, B/I_γ described in Theorem 10.24, R. S. Pierce [a] gave the first deep analysis of the structure of countable Boolean algebras. Pierce solved many known problems, among others the ones raised in P. R. Halmos [1963].

The investigations of Section 12 originated with the well-known theorem that the free Boolean algebras satisfy the Countable Chain Condition (see I. Reznikoff [1963] and A. Horn [1968]. In fact, this statement is true whenever a lattice structure is involved—see the beautiful proof in F. Galvin and B. Jónsson [1961].) The question arose whether this is so because the free Boolean algebras are free products of copies of the free Boolean algebra on one generator, which is finite and thus satisfies the Countable Chain Condition.

R. Balbes [1967] found another interesting property of free distributive lattices: Every subset S with the property $a \wedge b = c \wedge d$ for all $a,b,c,d \in S$, $a \neq b$, $c \neq d$ is finite, and for free Boolean algebras every such set is finite or countable. H. Lakser proved that if, for each such set S, we have $|S| < \aleph_0$ in each lattice, then $|S| \leq \aleph_0$ in a free $\{0, 1\}$-distributive product. This result has opened up a whole new field for investigation.

The development of the Amalgamation Property is described in B. Jónsson [1965]; probably B. Jónsson's work more than any other influence convinced the algebraists of the importance of this property. The Amalgamation Property for posets and lattices was noted by B. Jónsson [1956]; it was noted for Boolean algebras by A. Daigneault [1959].

As the reader may already have noticed, the conclusion of Theorem 13.21 can be based either on the Amalgamation Property, as suggested by B. Jónsson, or on the solution to the word problem for free products. (The solution to the word problem for distributive lattices is given in Corollary 12.6; this general method also works for lattices—see G. Grätzer, H. Lakser, and C. R. Platt [1970].) Neither of the two methods includes the other. For groups the first one works but not the second, whereas for semigroups it is the other way around.

There have been some further results on projective algebras. The theorem of R. Balbes describing finite projective distributive lattices has been extended to the countable case by R. Balbes and A. Horn [1970c]. R. Engelkind [1965] has shown that a subalgebra of a free Boolean algebra need not be projective.

The problem of how to prove the distributivity of a sublattice by imposing relations on the generators is discussed in many papers. For general lattices and three-generated sublattices, this problem was solved by O. Ore [1940]. It was observed that when the lattice is modular, the elements a, b, and c generate a distributive sublattice iff $a \wedge (b \vee c) = (a \wedge b) \vee (a \wedge c)$ (as is evident from Figure 5.7). Far-reaching generalizations of this result can be found in B. Jónsson [1955]; see also R. Musti and E. Buttafuoco [1956] and R. Balbes [1969].

The interdependence of the various "finiteness conditions" for distributive lattices was investigated in S. P. Avann [1964].

Boolean algebras with only the trivial automorphism were constructed by M. Katětov [1951], B. Jónsson [1951], and S. Rieger [1951]; Katětov's example has $2^{\aleph_0}$ elements; the others are very large.

The method used in establishing Lemma 10.22 was employed by B. Jónsson [1956] to construct "universal" lattices, that is, lattices containing a large class of lattices as sublattices (see also B. Jónsson [1960] and M. D. Morley and R. L. Vaught [1962]).

For background material in categories see B. Mitchell [1965].

R. Dedekind found the distributive identity by investigating ideals of number fields. Rings with a distributive lattice of ideals have been investigated by E. Noether [1927], L. Fuchs [1949] (who named such rings *arithmetical rings*), I. S. Cohen [1950], and Ch. U. Jensen [1963]. Equational classes of rings with distributive ideal lattices were considered in G. Michler and R. Wille [1970] and in H. Werner and R. Wille [1970]. E. A. Behrens [1960] and [1961] considered rings in which one-sided ideals form a distributive lattice. Rings with a distributive lattice of subrings were classified in P. A. Freĭdman [1967].

For groups the corresponding problem is much simpler: The subgroup lattice of a group G is distributive iff G is locally cyclic (see O. Ore [1937] and [1938]).

The distributivity of congruence lattices of lattices implies a number of important consequences. B. Jónsson [1967] discovered that many of these results hold for arbitrary universal algebras with distributive congruence lattices. His results have found applications that go far beyond lattice theory—they have already been applied to lattice-ordered algebras, closure algebras, nonassociative lattices, cylindric algebras, monadic algebras, lattices with pseudocomplementation, and primal algebras.

The foregoing examples show the central role played by distributive lattices in applications of the lattice concept.

PROBLEMS

28. Find short one-identity axioms characterizing Boolean algebras.
29. Is there a self-dual minimal set of identities defining Boolean algebras?
30. Characterize the automorphism groups of Boolean algebras.[9]
31. For a Boolean algebra B and cardinal number $\mathfrak{m} > |B|$, can a Boolean algebra C be found such that $|C| = \mathfrak{m}$ and the automorphism groups of B and C are isomorphic?[10]
32. Does the endomorphism semigroup determine a Boolean algebra up to isomorphism?[11]

---

[9] B. Jónsson has proved (unpublished result) that the finite automorphism groups of Boolean algebras are exactly the finite symmetric groups.

[10] B. Jónsson suggests starting the investigation of this problem with the one-element automorphism group.

[11] This problem was solved in the affirmative by B. Jónsson.

33. Characterize the poset $\mathscr{P}(L)$ of prime ideals of a distributive lattice.

34. Characterize $\mathscr{P}(L)$ under the additional assumption that L has a 0, a 1, or both. If L has 1, then every chain in $\mathscr{P}(L)$ has a supremum; if L has 0, then every chain in $\mathscr{P}(L)$ has an infimum. Are these the only additional conditions?

35. For which classes **K** of finite lattices can we claim that if $L \in$ **K** and L is nonmodular, then L contains a "minimal" $\mathfrak{N}_5$?

36. Characterize the congruence lattices of lattices. (Conjecture: distributive algebraic lattices).

37. Characterize the congruence lattices of sectionally complemented lattices. (Conjecture: distributive algebraic lattices).

38. Determine $f_{\mathbf{D}}(k, n)$.

39. Is a distributive lattice, in which every ideal is the intersection of maximal ideals, and dually, necessarily relatively complemented (see A. Monteiro [1947])?

40. Investigate the cardinalities of maximal chains in complete Boolean algebras.

41. What can be proved about the cardinalities of maximal chains in a complete and atomic Boolean algebra without the Generalized Continuum Hypothesis?

42. Compare the number and form of the terms in a normal $\vee$-representation and normal $\wedge$-representation of an element in a free $\{0, 1\}$-distributive product.

43. For a bounded distributive lattice L, let $F(L)$ and $F_n(L)$ denote the lattice of all functions and all n-ary functions on L, respectively, with the Congruence Substitution Property. To what extent do $F(L)$ and $F_n(L)$ determine the structure of L?

44. Characterize $F(L)$ and $F_n(L)$.

45. Describe the functions with the Congruence Substitution Property on unbounded distributive lattices.

46. Study problems 43 and 44 in the unbounded case.

47. For a lattice L, let $w(L)$ denote the smallest cardinal number such that $|S| < w(L)$ whenever S satisfies the following property: $a, b, c, d \in S$, $a \neq b$, and $c \neq d$ imply that $a \wedge b = c \wedge d$. For cardinals $\mathfrak{m}$ and $\mathfrak{n}$ ($\mathfrak{n} > 1$), let $Q(\mathfrak{m}, \mathfrak{n})$ denote the condition that whenever L_i, $i \in I$, are distributive lattices and L is the free distributive product, then $w(L_i) < \mathfrak{m}$ for all $i \in I$ and $|I| < \mathfrak{n}$ imply that $w(L) < \mathfrak{m}$. Describe the relation $Q(\mathfrak{m}, \mathfrak{n})$.

48. For cardinals $\mathfrak{m}$, $\mathfrak{n}$ ($\mathfrak{n} > 1$), define $R(\mathfrak{m}, \mathfrak{n})$ to hold if the free product of $\mathfrak{n}$ lattices each satisfying the $\mathfrak{m}$ chain condition satisfies the $\mathfrak{m}$ chain condition. Describe $R(\mathfrak{m}, \mathfrak{n})$. ($R(\aleph_1, \mathfrak{n})$ should always hold. Substitute G. Grätzer, H. Lakser, and C. R. Platt [1970] for Theorem 12.5).

49. Investigate $R(\mathfrak{m}, \mathfrak{n})$ for other classes of lattices (modular lattices, for example).

50. Is it possible to define formally the free $\{0, 1\}$-distributive product via the description given in Theorems 12.5 and 12.9?

51. Does the Amalgamation Property hold for the class of all modular lattices?[12]

52. Are there $2^{\aleph_0}$ equational classes of lattices for which the Amalgamation Property holds?

---

[12] In January 1971, B. Jónsson announced a negative solution to this problem.

CHAPTER 3

DISTRIBUTIVE LATTICES WITH PSEUDOCOMPLEMENTATION

14. Introduction and Stone Algebras

In this chapter we shall deal exclusively with pseudocomplemented distributive lattices. There are two concepts that we should be able to distinguish: a lattice, $\langle L; \wedge, \vee \rangle$, in which every element has a pseudo-complement, and an algebra, $\langle L; \wedge, \vee, *, 0, 1 \rangle$, where $\langle L; \wedge, \vee, 0, 1 \rangle$ is a bounded lattice and where, for every $a \in L$, the element a^* is a pseudocomplement of a. We shall call the former a *pseudocomplemented lattice* and the latter a *lattice with pseudocomplementation* (as an operation)—the same kind of distinction that we make between Boolean lattices and Boolean algebras. Thus, in the sense of the exercises following Section 12, a pseudocomplemented lattice is an algebra of type $\langle 2, 2 \rangle$, whereas a lattice with pseudocomplementation is an algebra of type $\langle 2, 2, 1, 0, 0 \rangle$. To see the difference in viewpoint, consider the finite distributive lattice of Figure 14.1. As a distributive lattice it has twenty-five sublattices and eight congruences; as a lattice with pseudocomplementation it has three sub-algebras and five congruences.

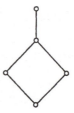

Figure 14.1

Thus, for a lattice with pseudocomplementation L, a *subalgebra* L_1 is a $\{0, 1\}$-sublattice of L closed under * (that is, $a \in L_1$ implies that $a^* \in L_1$). A *homomorphism* φ is a $\{0, 1\}$-homomorphism that also satisfies $(x\varphi)^* = x^*\varphi$. If there is any danger of confusion, we call such a homomorphism a *-homomorphism*. Similarly, a *congruence relation* Θ will have the Substitution Property also for *: $a \equiv b(\Theta)$ implies that $a^* \equiv b^*(\Theta)$.

A wide class of examples is provided by

THEOREM 1.   *Any complete lattice that satisfies the Join Infinite Distributive Identity (JID) is a pseudocomplemented distributive lattice.*

PROOF.   Let L be such a lattice. For $a \in L$, set

$$a^* = \bigvee (x \mid x \in L, a \wedge x = 0).$$

Then, by (JID),

$$a \wedge a^* = a \wedge \bigvee (x \mid a \wedge x = 0) = \bigvee (a \wedge x \mid a \wedge x = 0) = \bigvee 0 = 0.$$

Furthermore, if $a \wedge x = 0$, then $x \leq a^*$ by the definition of a^*; thus a^* is indeed the pseudocomplement of a.   ●

COROLLARY 2.   *Every distributive algebraic lattice is pseudocomplemented.*

PROOF.   Let L be a distributive algebraic lattice. By Theorem 9.13 and Lemma 11.1 we can assume that $L = I(S)$, where S is a distributive join-semilattice with 0. Let $I, I_j \in I(S)$ for $j \in J$. Then $I \wedge I_k \subseteq I \wedge \bigvee (I_j \mid j \in J)$ for any $k \in J$, and thus

$$\bigvee (I \wedge I_j \mid j \in J) \subseteq I \wedge \bigvee (I_j \mid j \in J).$$

To prove the reverse inclusion, let $a \in I \wedge \bigvee (I_j \mid j \in J)$, that is, $a \in I$, $a \in \bigvee (I_j \mid j \in J)$. The latter implies that $a \leq t_1 \vee \cdots \vee t_n$, where $t_1 \in I_{j_1}, \ldots, t_n \in I_{j_n}, j_1, \ldots, j_n \in J$. Thus $a \in I_{j_1} \vee \cdots \vee I_{j_n}$ and so, using the distributivity of L, we obtain

$$a \in I \wedge (I_{j_1} \vee \cdots \vee I_{j_n}) = (I \wedge I_{j_1}) \vee \cdots \vee (I \wedge I_{j_n}) \subseteq$$
$$\bigvee (I \wedge I_j \mid j \in J),$$

completing the proof of the (JID). The statement now follows from Theorem 1. ●

Thus, the lattice of all congruence relations of an arbitrary lattice and the lattice of all ideals of a distributive (semi-) lattice with zero are examples of pseudocomplemented distributive lattices. Note that for $I \in I(K)$,

$$I^* = \{x \mid x \in K, x \wedge i = 0 \text{ for all } i \in I\}.$$

Also, any finite distributive lattice is pseudocomplemented. Therefore, our investigations include all finite distributive lattices.

A model for distributive lattices with pseudocomplementation is the class of Boolean algebras, and the purpose of much of the research is to see how far they deviate from Boolean algebras. This purpose will be stated more precisely in the following paragraphs.

The first class of distributive lattices with pseudocomplementation, other than the class of Boolean algebras, to be examined in detail was the class of Stone algebras. A distributive lattice with pseudocomplementation L is called a *Stone algebra* iff it satisfies the *Stone identity*:

$$a^* \vee a^{**} = 1.$$

The corresponding pseudocomplemented lattice is called a *Stone lattice*. To understand the meaning of this identity, define the *skeleton* of L:

$$S(L) = \{a^* \mid a \in L\}.$$

The elements of $S(L)$ are called *skeletal*. L is *dense* if $S(L) = \{0, 1\}$. By Theorem 6.4, $\langle S(L); \wedge, \vee, *, 0, 1 \rangle$ is a Boolean algebra. For a Stone algebra L, $S(L)$ is a subalgebra of L:

THEOREM 3. *For a distributive lattice with pseudocomplementation L, the following conditions are equivalent:*

(i) *L is a Stone algebra.*

(ii) $(a \wedge b)^* = a^* \vee b^*$ for $a,b \in L$.

(iii) $a,b \in S(L)$ implies that $a \vee b \in S(L)$.

(iv) $S(L)$ is a subalgebra of L.

PROOF.   The proofs that (ii) implies (iii), that (iii) implies (iv), and that (iv) implies (i) are trivial. Now let L be a Stone algebra; we show that $a^* \vee b^*$ is the pseudocomplement of $a \wedge b$. Indeed, $(a \wedge b) \wedge (a^* \vee b^*) = (a \wedge b \wedge a^*) \vee (a \wedge b \wedge b^*) = 0 \vee 0 = 0$. If $(a \wedge b) \wedge x = 0$, then $(b \wedge x) \wedge a = 0$, and so $b \wedge x \le a^*$. Meeting both sides by a** yields $b \wedge x \wedge a^{**} \le a^* \wedge a^{**} = 0$; that is, $x \wedge a^{**} \wedge b = 0$, implying that $x \wedge a^{**} \le b^*$. By the Stone Identity, $a^* \vee a^{**} = 1$, and thus $x = x \wedge 1 = x \wedge (a^* \vee a^{**}) = (x \wedge a^*) \vee (x \wedge a^{**}) \le a^* \vee b^*$.   ●

This is already enough to yield the structure theorem for finite Stone algebras (G. Grätzer and E. T. Schmidt [1957b]):

COROLLARY 4.   *A finite distributive lattice is a Stone lattice iff it is the direct product of finite distributive dense lattices, that is, finite distributive lattices with only one atom.*

PROOF.   By Theorem 3, a Stone lattice L has a complemented element $a \notin \{0, 1\}$ iff $S(L) \neq \{0, 1\}$; thus the decomposition of Theorem 7.6 can be repeated until each factor L_i satisfies $S(L_i) = \{0, 1\}$. In a direct product, * is formed componentwise; therefore, all the L_i are Stone lattices. For a finite lattice K with $S(K) = \{0, 1\}$, the condition that K has one atom is equivalent to K being a Stone lattice.   ●

Another significant subset of a Stone algebra is the *dense set*

$$D(L) = \{a \mid a^* = 0\}.$$

The elements of $D(L)$ are called *dense*.

We can easily check that $D(L)$ is a dual ideal of L and that $1 \in D(L)$; thus $D(L)$ is a distributive lattice with 1. Since $a \vee a^* \in D(L)$ for every $a \in L$, we can interpret the identity

$$a = a^{**} \wedge (a \vee a^*)$$

to mean that every $a \in L$ can be represented in the form $a = b \wedge c$, where $b \in S(L)$, $c \in D(L)$. Such an interpretation correctly suggests that if we know $S(L)$ and $D(L)$ and the relationships between elements of $S(L)$

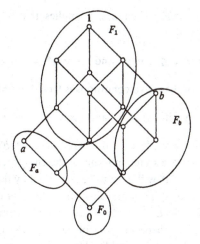

Figure 14.2

and $D(L)$, then we can describe L. The relationship is expressed by the homomorphism $\varphi(L)\colon S(L) \to \mathscr{D}(D(L))$ defined by

$$\varphi(L)\colon a \to \{x \mid x \in D(L),\ x \geq a^*\}.$$

THEOREM 5 (C. C. Chen and G. Grätzer [1969b]). *Let L be a Stone algebra. Then $S(L)$ is a Boolean algebra, $D(L)$ is a distributive lattice with* 1, *and $\varphi(L)$ is a $\{0, 1\}$-homomorphism of $S(L)$ into $\mathscr{D}(D(L))$. The triple $\langle S(L), D(L), \varphi(L)\rangle$ characterizes L up to isomorphism.*

PROOF.   The first statement is easily verified. For $a \in S(L)$, set

$$F_a = \{x \mid x^{**} = a\}.$$

The sets $\{F_a \mid a \in S(L)\}$ form a partition of L; for a simple example, see Figure 14.2. Obviously, $F_0 = \{0\}$ and $F_1 = D(L)$. The map $x \to x \vee a^*$ sends F_a into $F_1 = D(L)$; in fact, the map is an isomorphism between F_a and $a\varphi(L) \subseteq D(L)$. Thus $x \in F_a$ is completely determined by a and $x \vee a^* \in a\varphi(L)$—that is, by a pair $\langle a, z \rangle$, where $a \in S(L)$, $z \in a\varphi(L)$—and every such pair determines one and only one element of L. To complete our proof we have to show how the partial ordering on L can be determined by such pairs.

Let $x \in F_a$ and $y \in F_b$. Then $x \leq y$ implies that $x^{**} \leq y^{**}$, that is, $a \leq b$. Since $x \leq y$ iff

$$a \vee x \leq a \vee y \quad \text{and} \quad x \vee a^* \leq y \vee a^*$$

and since the first of these two conditions is trivial, we obtain:

$$x \leq y \quad \text{iff} \quad a \leq b \quad \text{and} \quad x \vee a^* \leq y \vee a^*.$$

Identifying x with $\langle x \vee a^*, a \rangle$ and y with $\langle y \vee b^*, b \rangle$, we see that the preceding conditions are stated in terms of the components of the ordered pairs, except that $y \vee a^*$ will have to be expressed by the triple.

Because $\varphi(L)$ is a $\{0, 1\}$-homomorphism and a^{**} is the complement of a^*, we conclude that $a^{**}\varphi(L)$ and $a^*\varphi(L)$ are complementary dual ideals of $D(L)$. Therefore, by Theorem 7.6, for any $z \in D(L)$, $[z)$ is the direct product of $[z \vee a^*)$ and $[z \vee a^{**})$. Thus, every z can be written in a unique fashion in the form $z = z(a^*) \wedge z(a^{**})$, where $z(a^*) \in a\varphi(L)$ and $z(a^{**}) \in a^*\varphi(L)$. Let $y\rho_a$ denote the element $(y\varphi(L))(a^*)$ and observe that ρ_a is expressed in terms of the triple. Finally, $y \vee a^* = y \vee b^* \vee a^* = (y\varphi(L)) \vee a^* = y\rho_a$. Thus for $u \in a\varphi(L)$ and $v \in b\varphi(L)$, we have $\langle u, a \rangle \leq \langle v, b \rangle$ iff $a \leq b$ and $u \leq v\rho_a$. ●

This theorem shows that for Stone algebras, the behavior of the skeleton and the dense set is decisive. This conclusion leads us to formulate the goal of research for Stone algebras and *for all* distributive lattices with pseudocomplementation as follows:

A problem for distributive lattices with pseudocomplementation is considered solved if it can be reduced to two problems: one for Boolean algebras and one for distributive lattices with 1. We shall see examples in which this program works and others in which it does not.

Exercises

1. Show that every bounded chain is a pseudocomplemented distributive lattice.
2. Let L be a lattice with 1. Adjoin a new zero 0 to L: $L_1 = L \cup \{0\}$, $0 < x$ for all $x \in L$. Show that L_1 is a pseudocomplemented lattice.
3. Call a lattice with 0 *dense* if 0 is meet-irreducible. Show that every bounded dense lattice K is pseudocomplemented and that every such lattice can be constructed by the method of exercise 2 with $L = D(K)$.

4. Find an example of a complete distributive lattice L that is not pseudo-complemented.

5. Prove that if L is a complete Stone lattice, then so is $I(L)$. (Hint: $I^* = (a]$, where $a = \bigwedge (x^* \mid x \in I)$.)

6. Show that a distributive pseudocomplemented lattice is a Stone lattice iff $(a \lor b)^{**} = a^{**} \lor b^{**}$ for $a,b \in L$.

7. Find a small set of identities characterizing Stone algebras.

8. Let L be a Stone algebra. Show that $S(L)$ is a retract of L.

9. Let L be a Stone algebra, $a,b \in S(L)$, and $a \le b$. Prove that $x \to (x \lor a^*) \land b$ embeds F_a into F_b.

10. Let B be a Boolean lattice. Define $B^{[2]} \subseteq B^2$ by $\langle a, b \rangle \in B^{[2]}$ iff $a \le b$. Verify that $B^{[2]}$ is a sublattice of B^2 but not a subalgebra of B^2. Show that $B^{[2]}$ is a Stone lattice.

Exercises 11–23 are from C. C. Chen and G. Grätzer [1969b]. Exercises 24 and 25 are from C. C. Chen and G. Grätzer [1969c].

Let B be a Boolean algebra, let D be a distributive lattice with 1, and let φ be a $\{0, 1\}$-homomorphism of B into $\mathscr{D}(D)$. Set $L = \{\langle x, a \rangle \mid a \in B, x \in a\varphi\}$ and define $\langle x, a \rangle \le \langle y, b \rangle$ iff $a \le b$ and $x \le y\rho_a$, where $[y\rho_a] = a\varphi \land [y)$.

11. Verify the following formulas:
    (i) If $a \in B$ and $d \in D$, then $d\rho_a = d$ iff $d = a\varphi$.
    (ii) $d\rho_a \ge d$ for $a \in B, d \in D$.
    (iii) $d\rho_a \land d\rho_{a'} = d$ for $a \in B, d \in D$ (where a' is the complement of a in B).
    (iv) $\rho_a\rho_b = \rho_{a \land b}$ for $a,b \in C$.

12. Prove that:
    (i) $d\rho_a \land d\rho_b = d\rho_{a \lor b}$ for $a,b \in B, d \in D$.
    (ii) $d\rho_{a \land b} = d\rho_a \lor d\rho_b$ for $a,b \in B, d \in D$.

13. Show that L is a poset under the given partial ordering.

14. Let $\langle x, a \rangle, \langle y, b \rangle \in L$. Verify that $\langle x, a \rangle \land \langle y, b \rangle = \langle x\rho_b \land y\rho_a, a \land b \rangle$.

*15. Show that $\langle x, a \rangle \lor \langle y, b \rangle = \langle (x\rho_{b'} \land y) \lor (x \land y\rho_{a'}), a \lor b \rangle$.

16. Let $\langle x, a \rangle, \langle y, b \rangle$, and $\langle z, c \rangle \in L$;

$$A = (\langle x, a \rangle \land \langle y, b \rangle) \lor \langle z, c \rangle,$$

and

$$B = (\langle x, a \rangle \lor \langle z, c \rangle) \land (\langle y, b \rangle \lor \langle z, c \rangle).$$

Compute A; show that $B = \langle d, (a \lor c) \land (b \lor c) \rangle$, where $d = d_0 \lor d_1 \lor d_2 \lor d_3$, and $d_0 = x\rho_{b \land c'} \land y\rho_{a \land c'} \land z, d_1 = x\rho_{b \land c'} \land y\rho_{a \land c} \land z, d_2 = x\rho_{b \lor c} \land y\rho_{a \land c'} \land z$, and $d_3 = x\rho_{b \lor c} \land y\rho_{b \lor c} \land z\rho_{a' \lor b'}$.

17. Show that $d_0 \ge d_1$ and $d_0 \ge d_2$; therefore, $d = d_0 \lor d_3$.

18. Show that L is distributive.

19. Show that L is a Stone lattice.

20. Identify $a \in B$ with $\langle 1, b \rangle$ and $d \in D$ with $\langle d, 1 \rangle$. Verify that then $S(L) = B, D(L) = D$, and $\varphi(L) = \varphi$. In other words, we have proved the Construction Theorem of C. C. Chen and G. Grätzer [1969b]: *Given a*

Boolean algebra B, a distributive lattice D with 1, and a {0, 1}-homomorphism $\varphi: B \to \mathscr{D}(D)$, *there exists a Stone algebra L whose triple is* $\langle B, D, \varphi \rangle$.

21. Describe isomorphisms and homomorphisms of Stone algebras in terms of triples.

22. Describe subalgebras of Stone algebras in terms of triples.

23. For a given Boolean algebra B with more than one ele..ient and distributive lattice D with 1, construct a Stone algebra with $S(L) \cong B$, $D(L) \cong D$. ($S(L)$ and $D(L)$ are independent.)

24. Show that a Stone algebra L is complete if $S(L)$ and $D(L)$ are complete.

*25. Characterize the completeness of Stone algebras in terms of triples.

26. Let L be a distributive lattice. Prove that L is a Stone lattice iff a unary operation * can be defined on L that satisfies the following implication:

$$a \wedge b \leq c \leq a \vee b \quad \text{implies that} \quad c \wedge b^* \leq a \leq c^{**} \vee b^*$$

(K. M. Koh).

15. Identities and Congruences

Since one of our objectives in this chapter is to learn how to compute with *, we might as well start by collecting some important rules of the arithmetic of the operations *, $\wedge$, $\vee$.

THEOREM 1. *Let L be a pseudocomplemented distributive lattice,* $S(L) = \{a^* \mid a \in L\}$, *and* $D(L) = \{a \mid a^* = 0\}$. *Then for* $a, b \in L$:

(i) $a \wedge a^* = 0$.

(ii) $a \leq b$ *implies that* $a^* \geq b^*$.

(iii) $a \leq a^{**}$.

(iv) $a^* = a^{***}$.

(v) $(a \vee b)^* = a^* \wedge b^*$.

(vi) $(a \wedge b)^{**} = a^{**} \wedge b^{**}$.

(vii) $a \wedge b = 0$ *iff* $a^{**} \wedge b^{**} = 0$.

(viii) $a \wedge (a \wedge b)^* = a \wedge b^*$.

(ix) $0^* = 1$ *and* $1^* = 0$.

(x) $a \in S(L)$ *iff* $a = a^{**}$.

(xi) $a, b \in S(L)$ *implies that* $a \wedge b \in S(L)$.

(xii) $\sup_{S(L)} \{a, b\} = (a \vee b)^{**} = (a^* \wedge b^*)^*$.

(xiii) $0, 1 \in S(L)$, $1 \in D(L)$, *and* $S(L) \cap D(L) = \{1\}$.

(xiv) $a, b \in D(L)$ *implies that* $a \wedge b \in D(L)$.

(xv) $a \in D(L)$ *and* $a \leq b$ *imply that* $b \in D(L)$.

(xvi) $a \lor a^* \in D(L)$.

(xvii) $x \to x^{**}$ *is a meet-homomorphism of L onto S(L)*.

PROOF. (i), (ii), (iii), and (iv) were proved in Section 6. To show (v), observe that $(a \lor b) \land (a^* \land b^*) = (a \land a^* \land b^*) \lor (b \land a^* \land b^*) = 0 \lor 0 = 0$. Now $(a \lor b) \land x = 0$ implies that $(a \land x) \lor (b \land x) = 0$; therefore, $a \land x = 0$ and $b \land x = 0$, and so $x \le a^*, x \le b^*$, thus $x \le a^* \land b^*$. Note that x^{**} is the smallest element of $S(L)$ containing x. By Formula (5) of Section 6, $a^{**} \land b^{**} \in S(L)$, and it is obviously the smallest element of $S(L)$ containing $a \land b$, hence (vi); (vii) is an immediate consequence of (vi) and (iii); (viii)–(xv) and (xvii) are either trivial or known from Section 6; (xvi) follows directly from (v). ●

The Stone identity has a nice generalization to an identity in n variables, which is attributed to K. B. Lee [1970]:

(L_n) $(x_1 \land \cdots \land x_n)^* \lor (x_1^* \land \cdots \land x_n)^* \lor \cdots \lor (x_1 \land \cdots \land x_n^*)^* = 1$.

Note that (L_1) is $x_1^* \lor x_1^{**} = 1$, the Stone identity. For $n = 2$ and $n = 3$ we obtain

(L_2) $(x_1 \land x_2)^* \lor (x_1^* \land x_2)^* \lor (x_1 \land x_2^*)^* = 1$,

(L_3) $(x_1 \land x_2 \land x_3)^* \lor (x_1^* \land x_2 \land x_3)^* \lor (x_1 \land x_2^* \land x_3)^* \lor$

$$(x_1 \land x_2 \land x_3^*)^* = 1.$$

The larger n is, the harder (L_n) is to work with. The following lemma (G. Grätzer and H. Lakser [1969b]) is sometimes useful:

LEMMA 2. *Let L be a distributive lattice with pseudocomplementation. Then the following conditions are equivalent:*

(i) *L satisfies the identity* (L_n).

(ii) *L satisfies the implication:*

$$x_i \land x_j = 0, i \ne j, i, j = 0, \ldots, n, \text{ imply that}$$

$$x_0^* \lor x_1^* \lor \cdots \lor x_n^* = 1.$$

(iii) *L satisfies the implication:*

$$x_i \land x_j = 0, i \ne j, i, j = 0, \ldots, n, x_0, \ldots, x_n \in S(L)$$

$$\text{imply that } x_0^* \lor x_1^* \lor \cdots \lor x_n^* = 1.$$

REMARK. For $n = 1$, (ii) is: $x_0 \wedge x_1 = 0$ implies that $x_0^* \vee x_1^* = 1$. For $n = 2$, (ii) is: $x_0 \wedge x_1 = x_0 \wedge x_2 = x_1 \wedge x_2 = 0$ imply that $x_0^* \vee x_1^* \vee x_2^* = 1$.

PROOF.

(i) *implies* (ii). Let $x_i \wedge x_j = 0$ for $i \neq j$, $i, j = 0, \ldots, n$. Then for $i \neq j$, $x_i \leq x_j^*$; therefore, $x_0 \leq x_1^* \wedge \cdots \wedge x_n^*, x_1 \leq x_1^{**} \wedge x_2^* \wedge \cdots \wedge x_n^*$, $\ldots, x_n \leq x_1^* \wedge x_2^* \wedge \cdots \wedge x_n^{**}$. Thus by applying (L_n) to $x_1^*, \ldots, x_n^*$ we obtain

$$x_0^* \vee x_1^* \vee \cdots \vee x_n^* \geq (x_1^* \wedge \cdots \wedge x_n^*)^* \vee (x_1^{**} \wedge \cdots \wedge x_n^*)^* \vee \cdots$$
$$\vee (x_1^* \wedge \cdots \wedge x_n^{**})^* = 1. \quad \blacksquare$$

(ii) *implies* (iii), by specialization. $\quad \blacksquare$

(iii) *implies* (i). Set $y_0 = x_1 \wedge \cdots \wedge x_n, y_1 = x_1^* \wedge \cdots \wedge x_n, \ldots, y_n = x_1 \wedge \cdots \wedge x_n^*$. Then $y_i \wedge y_j = 0$ for $i \neq j$, and so, by Theorem 1(vii), we have $y_i^{**} \wedge y_j^{**} = 0$, for $i \neq j$, and $y_0^{**}, \ldots, y_n^{**} \in S(L)$. We can thus apply (iii) to the elements $y_0^{**}, \ldots, y_n^{**}$ to obtain $y_0^{***} \vee y_1^{***} \vee \cdots \vee y_n^{***} = 1$. This reduces to (L_n) by Theorem 1(iv). $\quad \bullet$

As an immediate application of Lemma 2 we can check whether (L_n) holds in certain examples. Let $\mathfrak{C}_n$ denote the 2^n-element Boolean lattice with a new 1 (see $\mathfrak{C}_0$, $\mathfrak{C}_1$, $\mathfrak{C}_2$, and $\mathfrak{C}_3$ in Figure 15.1). More formally, $\mathfrak{C}_n$ is a bounded lattice with 0 and 1 and an element e such that $[0, e]$ is a 2^n-element Boolean lattice, $e \prec 1$, and $\mathfrak{C}_n = [0, e] \cup \{1\}$. It is easily seen that $\mathfrak{C}_n$ is pseudocomplemented; in fact, $1^* = 0, 0^* = 1$, and for every $x \in [0, e]$, $x \neq 0$, the pseudocomplement of x in $\mathfrak{C}_n$ is the relative complement of x in $[0, e]$.

LEMMA 3. *Let m be a nonnegative integer and let n be a positive integer. Then (L_n) holds in $\mathfrak{C}_m$ iff $m \leq n$.*

PROOF. Note that $S(\mathfrak{C}_m) \cong [0, e]$, which is a 2^m-element Boolean lattice having m atoms $p_1, \ldots, p_m$. Thus if $m \leq n$, then (since $S(\mathfrak{C}_m)$ does not have $n + 1$ pairwise disjoint nonzero elements) Lemma 2(iii) holds in $\mathfrak{C}_m$ trivially, and so by Lemma 2, (L_n) holds in $\mathfrak{C}_m$. If $n < m$, then $p_1, \ldots, p_{n+1}$ satisfy $p_i \wedge p_j = 0$ for $i \neq j$, $p_i \in S(\mathfrak{C}_m)$, but $p_1^* \vee \cdots \vee p_{n+1}^* \leq e < 1$. Thus Lemma 2(iii), and therefore (L_n) fails in $\mathfrak{C}_m$. $\quad \bullet$

Now we turn our attention to congruence relations. The intimate relation between congruences and prime ideals that we observed in

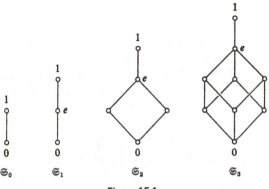

Figure 15.1

distributive lattices prevails for distributive lattices with pseudocomplementation. A new feature is the important role played by minimal prime ideals. A prime ideal P is called *minimal* if there is no prime ideal Q with $Q \subset P$.

LEMMA 4.   *Let L be a lattice with 0. Then every prime ideal contains a minimal prime ideal.*

PROOF.   Let P be a prime ideal of L and let $\mathscr{X}$ denote the set of all prime ideals Q contained in P. Then $\mathscr{X}$ is not void since $P \in \mathscr{X}$. If C is a chain in $\mathscr{X}$ and $Q = \bigcap (X \mid X \in C)$, then Q is nonvoid because $0 \in Q$ and Q is an ideal; in fact, Q is prime. Indeed, if $a \wedge b \in Q$ for some $a,b \in L$, then $a \wedge b \in X$ for all $X \in C$; since X is prime, either $a \in X$ or $b \in X$. Thus either $Q = \bigcap (X \mid a \in X)$ or $Q = \bigcap (X \mid b \in X)$, proving that a or $b \in Q$. Therefore, we can apply to $\mathscr{X}$ the dual form of Zorn's Lemma (which is also an equivalent form of the Axiom of Choice) to conclude the existence of a minimal member of $\mathscr{X}$.   ●

The following characterization of minimal prime ideals is quite useful.

LEMMA 5.   *Let L be a pseudocomplemented distributive lattice and let P be a prime ideal of L. Then the following four conditions are equivalent:*
  (i) *P is minimal.*
  (ii) *$x \in P$ implies that $x^* \notin P$.*
  (iii) *$x \in P$ implies that $x^{**} \in P$.*
  (iv) *$P \cap D(L) = \varnothing$.*

PROOF.

(i) *implies* (ii). Let P be minimal and let (ii) fail, that is, $a^* \in P$ for some $a \in P$. Let $D = (L - P) \vee [a)$. We claim that $0 \notin D$. Indeed, if $0 \in D$, then $0 = q \wedge a$ for some $q \in L - P$, which implies that $q \leq a^* \in P$, a contradiction. Thus by Theorem 7.15, there exists a prime ideal Q disjoint to D. Then $Q \subseteq P$ since $Q \cap (L - P) = \varnothing$, and $Q \neq P$ since $a \notin Q$, contradicting the minimality of P. ▶

(ii) *implies* (iii). Indeed, $x^* \wedge x^{**} = 0 \in P$ for any $x \in L$; thus if $x \in P$, then by (ii), $x^* \notin P$, implying that $x^{**} \in P$. ▶

(iii) *implies* (iv). If $a \in P \cap D(L)$ for some $a \in L$, then $a^{**} = 1 \notin P$, a contradiction to (iii). Thus $P \cap D(L) = \varnothing$. ▶

(iv) *implies* (i). If P is not minimal, then $Q \subset P$ for some prime ideal Q of L. Let $x \in P - Q$. Then $x \wedge x^* = 0 \in Q$ and $x \notin Q$; therefore $x^* \in Q \subset P$, which implies that $x \vee x^* \in P$. By Theorem 1(xvi), $x \vee x^* \in D(L)$; thus we obtain $x \vee x^* \in P \cap D(L)$, contradicting (iv.) ●

The importance of minimal ideals for Stone algebras was observed in G. Grätzer and E. T. Schmidt [1957b]; their importance in describing homomorphisms onto $\mathfrak{S}_1$ was illustrated by R. Balbes and G. Grätzer [a]. The next lemma, which (in one direction) is due to K. B. Lee [1970], generalizes this observation to arbitrary $\mathfrak{S}_n$.

LEMMA 6. *Let L be a distributive lattice with pseudocomplementation. The homomorphisms of L onto $\mathfrak{S}_n$ are in one-to-one correspondence with sequences $\langle P, Q_1, \ldots, Q_n \rangle$ of prime ideals of L, where $Q_1, \ldots, Q_n$ are all the distinct minimal prime ideals contained in P.*

PROOF. If φ is a homomorphism of L onto $\mathfrak{S}_n$, $p_1, \ldots, p_n$ are the atoms of $\mathfrak{S}_n$, and $e = p_1 \vee \cdots \vee p_n$, then we set $P = \{x \mid x \in L, x\varphi \leq e\}$, $Q_i = \{x \mid x \in L, x\varphi \leq p_i^*\}$, $i = 1, \ldots, n$. Clearly, $P, Q_1, \ldots, Q_n$ are prime ideals, and $Q_1, \ldots, Q_n$ satisfy Lemma 5(iv); thus, $Q_1, \ldots, Q_n$ are minimal prime ideals contained in P. Let R be a minimal prime ideal of L contained in P and distinct from all the Q_i. Then there exist $a_1 \in R - Q_1, \ldots, a_n \in R - Q_n$, and so $a = a_1 \vee \cdots \vee a_n$ satisfies $a \in R - Q_i$ for all $i = 1, \ldots, n$. Therefore, $a\varphi \not\leq p_i^*$ for all i. This implies that $a\varphi = e$ or 1. Thus $(a\varphi)^{**} = 1$ and so $a^{**}\varphi = 1$, implying that $a^{**} \notin P$, contradicting $a^{**} \in R$, which holds by Lemma 5(iii).

Conversely, let $P, Q_1, \ldots, Q_n$ be given as in the lemma and let us define the map φ by

$$a\varphi = \begin{cases} 1 & \text{if } a \notin P, \\ \bigvee (p_i \mid a \notin Q_i) & \text{for } a \in P. \end{cases}$$

We know from Section 9 that φ is a lattice homomorphism. To verify that $(a\varphi)^* = a^*\varphi$, first assume that $a \notin P$; in this case $a \wedge a^* \in Q_i$, and thus $a^* \in Q_i$ for $i = 1, \ldots, n$; that is, $a^* \in Q_1 \cap \cdots \cap Q_n$. By definition, then, $a^*\varphi = 0$. Also, $a\varphi = 1$, and so $(a\varphi)^* = 1^* = 0 = a^*\varphi$. Second, let $a \in Q_1 \cap \cdots \cap Q_n$. If $a^* \in P$, then, just as in the first step of Lemma 5, we conclude that there is a minimal prime ideal Q satisfying $a \notin Q \subseteq P$, a contradiction. Therefore, $a^* \notin P$ and so $(a\varphi)^* = 0^* = 1 = a^*\varphi$. Third, let $a \in P - Q_i$ for some i, $1 \le i \le n$. Then $a\varphi \ne 0$ and $a\varphi \in [0, e]$; thus $(a\varphi)^*$ is the complement of $a\varphi$ in $[0, e]$. Therefore,

$$(a\varphi)^* = (\bigvee (p_i \mid a \notin Q_i))^* = \bigvee (p_i \mid a \in Q_i) = \bigvee (p_i \mid a^* \notin Q_i) = a^*\varphi.$$

Finally, we have to show that φ is onto. Obviously, $0, 1 \in L\varphi$. To finish the proof it suffices to show that $p_1^*, \ldots, p_n^* \in L\varphi$. If we can choose $a \in Q_i$ such that $a \notin Q_j$ for all $j \ne i$, then $a\varphi = p_i^*$ as desired. Otherwise, for some Q_i, say Q_1, we have $Q_1 \subseteq Q_2 \cup \cdots \cup Q_n$. Choose a minimal set $I \subseteq \{2, \ldots, n\}$ such that $Q_1 \subseteq \bigcup (Q_i \mid i \in I)$. Then for each $i \in I$ there is an $a_i \in Q_1 - \bigcup (Q_j \mid j \in I - \{i\})$. Let $a = \bigvee (a_i \mid i \in I)$. Then $a \in Q_1$, and so $a \in Q_j$ for some $j \in I$; therefore, $a_i \in Q_j$ for all $i \in I$, contradicting $a_i \notin Q_j$ for $i \ne j$. ●

A description of all congruence relations of a distributive lattice with pseudocomplementation has been given by H. Lakser [a]. His result, which is the next theorem, is an interesting application of the principle stated at the end of Section 14.

Let L be a distributive lattice with pseudocomplementation. For a congruence relation Θ of L, let Θ_S and Θ_D denote the restrictions of Θ to $S(L)$ and $D(L)$, respectively. Obviously, Θ_D is a congruence relation of $D(L)$. In $S(L)$ the operations are $x \wedge y, x \vee y = (x^* \wedge y^*)^*$, and $*$; therefore, Θ_S is clearly a congruence relation of $S(L)$. Thus $\langle \Theta_S, \Theta_D \rangle \in C(S(L)) \times C(D(L))$. An arbitrary pair $\langle \Phi, \Psi \rangle \in C(S(L)) \times C(D(L))$ will be called a *congruence pair* if $a \in S(L), u \in D(L), u \ge a$, and $a \equiv 1(\Phi)$ imply that $u \equiv 1(\Psi)$.

THEOREM 7. *Let L be a distributive lattice with pseudocomplementation. Then every congruence relation Θ of L determines a congruence pair $\langle \Theta_S, \Theta_D \rangle$. Conversely, every congruence pair $\langle \Theta_1, \Theta_2 \rangle$ uniquely determines a congruence relation Θ on L with $\Theta_S = \Theta_1$ and $\Theta_D = \Theta_2$ by the following rule:*

$$x \equiv y(\Theta) \quad \textit{iff (i)} \quad x^* \equiv y^*(\Theta_1)$$

$$\textit{and (ii)} \; x \vee u \equiv y \vee u(\Theta_2) \textit{ for all } u \in D(L).$$

PROOF. The first statement is obvious. Let Θ be a congruence of L, $x,y \in L$, $x \equiv y(\Theta)$. By Theorem 1(i) and (iii), $x = x^{**} \wedge (x \vee x^*)$, $y = y^{**} \wedge (y \vee y^*)$, and $x^{**} \equiv y^{**}(\Theta_S)$, $x \vee x^* \equiv y \vee y^*(\Theta_D)$; thus Θ_S and Θ_D do indeed determine Θ.

Let $\langle \Theta_1, \Theta_2 \rangle$ be a congruence pair and let Θ be defined by (i) and (ii). Θ is obviously an equivalence relation. Theorem 1(v) implies the Substitution Property for $\vee$; for $\wedge$ it follows from Theorem 15.1(iv) and (vi). To show the Substitution Property for *, let $x \equiv y(\Theta)$. Then by (i), $x^* \equiv y^*(\Theta_1)$, and thus $x^{**} \equiv y^{**}(\Theta_1)$, which is (i) for x^* and y^*. Since $x^* \equiv y^*(\Theta_1)$ and $S(L)$ is Boolean, by Corollary 9.4 and Theorem 9.7 there is an $a \in S(L)$ such that $a \equiv 1(\Theta_1)$ and $x^* \wedge a \equiv y^* \wedge a(\Theta_1)$. Thus for any $u \in D(L)$ we obtain $u \vee a \equiv 1(\Theta_2)$ by the definition of the congruence pair, and so

$$x^* \vee u \equiv (x^* \vee u) \wedge (a \vee u) = (x^* \wedge a) \vee u \equiv (y^* \wedge a) \vee u$$

$$= (y^* \vee u) \wedge (a \vee u) \equiv y^* \vee u(\Theta_2),$$

proving (ii) for x^* and y^*. Therefore, Θ is a congruence relation.

For $x,y \in S(L)$, $x \equiv y(\Theta)$ iff $x^* \equiv y^*(\Theta_1)$ (since (ii) is trivial), and so $x \equiv y(\Theta_S)$ iff $x \equiv y(\Theta_1)$, that is, $\Theta_S = \Theta_1$. For $x,y \in D(L)$, (i) is trivial, and thus $x \equiv y(\Theta)$ iff, for all $u \in D(L)$, we have $x \vee u \equiv y \vee u(\Theta_2)$, which is equivalent to $x \equiv y(\Theta_2)$, and so $\Theta_2 = \Theta_D$. ●

We shall apply this description of congruences to establish an important property of pseudocomplemented distributive lattices, namely, the analogue of Theorem 9.6. To facilitate this we prove (a part of) the so-called Second Isomorphism Theorem of Algebra (see, for example, G. Grätzer [1968], Theorem 11.4). The word "algebra" in this lemma could be replaced by "lattice" or "lattice with pseudocomplementation."

LEMMA 8. *Let L be an algebra and let Θ be a congruence relation of L. For any congruence Φ of L such that $\Phi \geq \Theta$, define the relation Φ/Θ on L/Θ by*

$$[x]\Theta \equiv [y]\Theta(\Phi/\Theta) \quad \textit{iff } x \equiv y(\Phi).$$

Then Φ/Θ is a congruence of L/Θ. Conversely, every congruence Ψ of L/Θ can be (uniquely) represented in the form $\Psi = \Phi/\Theta$ for some congruence $\Phi \geq \Theta$.

PROOF.   We have to prove that Φ/Θ (i) is well-defined, (ii) is an equivalence relation, and (iii) has the Substitution Property. To represent Ψ, define Φ by

$$x \equiv y(\Phi) \quad \text{iff } [x]\Theta \equiv [y]\Theta(\Psi).$$

Again, we have to verify that Φ is a congruence. $\Phi/\Theta = \Psi$ follows from the definition of Φ. The trivial details are left to the reader.   ●

DEFINITION 9.   *A class* **K** *of algebras is said to have the* Congruence Extension Property *if, for* $A,B \in$ **K** *with A a subalgebra of B and* Θ *a congruence of A, there exists a congruence* Φ *on B such that* $\Phi_A = \Theta$, *that is,* Φ *restricted to A is* Θ.

Using this terminology, Theorem 9.6 states that the class of distributive lattices **D** has the Congruence Extension Property.

THEOREM 10 (G. Grätzer and H. Lakser [a]).   *The class of all distributive lattices with pseudocomplementation enjoys the Congruence Extension Property.*

PROOF.   Let L and K be distributive lattices with pseudocomplementation, let L be a subalgebra of K, and let Θ be a congruence of L given by the congruence pair $\langle \Theta_1, \Theta_2 \rangle$. It is clear from Theorem 7 that we need only show the existence of a congruence pair $\langle \Phi_1, \Phi_2 \rangle$ of K such that $(\Phi_1)_{S(L)} = \Theta_1$ and $(\Phi_2)_{D(L)} = \Theta_2$.

Let $J_L = [1]\Theta_1$ and put $J_K = [J_L)$, the dual ideal generated by J_L in $S(K)$. Then Φ_1 can be defined as the congruence of $S(K)$ associated with J_K, that is, $[1]\Phi_1 = J_K$. Set $I = \{i \mid i \in D(K), i \geq u \text{ for some } u \in J_L\}$. Then I is a dual ideal of $D(K)$; in fact, $I = [J_K) \cap D(K)$. By the definition of congruence pair, we have to find a congruence Φ_2 on $D(K)$ such that $(\Phi_2)_{D(L)} = \Theta_2$ and $[1]\Phi_2 \supseteq I$. Note that I has the following property:

If $u \in I, v \in D(L)$, and $v \leq u$, then there exists a $v_1 \in D(L)$, $v_1 \leq u$ such that $v_1 \in [1]\Theta_2$.

Indeed, $u \in I$ means that $u \geq x$ for some $x \in J_L$, and thus $v_1 = v \vee x$ will do the trick.

Summarizing, to complete the proof it suffices to prove the following statement:

Let A and B be distributive lattices with 1, A a $\{1\}$-sublattice of B, Θ a congruence of A, and I a dual ideal of B satisfying the condition:

If $u \in I$, $v \in A$, and $v \leq u$, then $v_1 \leq u$ for some $v_1 \in [1]\Theta$.

Then there exists a congruence relation Φ on B satisfying $\Phi_A = \Theta$ and $[1]\Phi \supseteq I$.

To prove this statement, consider $\Theta[I]$ defined by the dual of Corollary 9.4. If $a,b \in A$ and $a \equiv b(\Theta[I])$, then $a \wedge b = (a \vee b) \wedge i$ for some $i \in I$. Thus by our assumption on I, there is an $i_1 \in [1]\Theta$ such that $i_1 \leq i$. Therefore, $a \wedge b = (a \vee b) \wedge i \geq (a \vee b) \wedge i_1 \equiv (a \vee b) \wedge 1 = a \vee b(\Theta)$, and so $a \equiv b(\Theta)$. Having shown that $(\Theta[I])_A \leq \Theta$, we can form $A/\Theta[I]_A$, $B/\Theta[I]$, and $\Theta[I]_A/\Theta$. By Theorem 9.6 there exists a congruence Ψ on $B/\Theta[I]$ such that Ψ restricted to $A/\Theta[I]_A$ is $\Theta[I]_A/\Theta$. By Lemma 8 there is a unique congruence Φ on B such that $\Phi/\Theta[I] = \Psi$ and $\Phi \geq \Theta[I]$. Obviously, Φ satisfies the requirements. ●

Exercises

1. List the rules in Theorem 1 that hold for all pseudocomplemented lattices.

2. Let L be a distributive lattice in which every prime ideal contains a minimal prime ideal. Prove that L has 0.

3. Let A and B be dual ideals of a pseudocomplemented distributive lattice L. Verify that $A \cap S(L)$ and $B \cap S(L)$ are dual ideals of $S(L)$ and $(A \vee B) \cap S(L) = (A \cap S(L)) \vee (B \cap S(L))$.

4. For a pseudocomplemented distributive lattice L, define the relation R by: $x \equiv y(R)$ iff $x^* = y^*$. Show that R is a congruence on L and $L/R \cong S(L)$.

5. Show that in a Stone algebra every prime ideal contains exactly one minimal prime ideal.

6. Prove that a prime ideal P of a Stone algebra L is minimal iff $P = (P \cap S(L)]_L$.

7. Let Θ be a principal congruence relation of a distributive lattice L with pseudocomplementation. Show that $\Theta_{D(L)}$ need not be principal.

8. Conclude from Lemma 8 that $[\Theta)$ of $C(L)$ is isomorphic to $C(L/\Theta)$.

9. Does Theorem 10 imply the Congruence Extension Property for **D**?

*10. Find a proof of Theorem 10 that does not make use of the Congruence Extension Property for **D** (G. Grätzer and H. Lakser [a]).

11. Show that a distributive lattice with pseudocomplementation is a Stone algebra iff every prime ideal contains exactly one minimal prime ideal (G. Grätzer and E. T. Schmidt [1957b]).

*12. Prove that a poset Q is isomorphic to the poset of all prime ideals of a
Stone algebra iff (i) every element of Q contains exactly one minimal
element, and (ii) for every minimal element m of Q, the poset $\{x \mid x > m,$
$x \in Q\}$ is isomorphic to the poset of all prime ideals of some distributive
lattice with 1 (C. C. Chen and G. Grätzer [1969c]).

13. A lattice L is a *relative Stone lattice* if every interval of L is a Stone
lattice. Show that the Stone lattice L is a relative Stone lattice iff $D(L)$ is a
relative Stone lattice (C. C. Chen and G. Grätzer [1969c]).

14. Find a complete Boolean lattice B such that $\mathfrak{S}_1$ is a subalgebra of $I(B)$.

Exercises 15 ànd 16 are from B. Banaschewski [1970].

15. Let **K** be an equational class of algebras in which every algebra can be
embedded in an injective algebra. Show that **K** has the Congruence
Extension Property.

16. Let **K** be the same as in exercise 15. Show that **K** has the Amalgamation
Property.

17. Find a direct proof of the Congruence Extension Property for Stone
algebras.

18. For a class **K** of algebras, let H(K), S(K), and I(K) denote the classes of
homomorphic images of algebras in **K**, subalgebras of algebras in **K**, and
algebras isomorphic to algebras in **K**, respectively. Show that if **K** has the
Congruence Extension Property, then HS(K) = ISH(K).

19. Show that the congruence pairs form a sublattice of $C(S(L)) \times$
$C(D(L))$ (H. Lakser [a]).

20. Let L be a distributive lattice with 0 and 1. For an ideal I of L, we set
$I^* = \{x \mid x \wedge i = 0 \text{ for all } i \in I\}$. Let M be a prime ideal of L. Show that
M is a minimal prime ideal iff $x \in M$ implies that $(x]^* \nsubseteq M$ (T. P.
Speed).

21. Let L_1 and L_2 be Stone algebras and let φ be a $\{0, 1\}$-homomorphism of
L_1 into L_2. Show that φ is an algebra homomorphism iff $D(L_1)\varphi \subseteq$
$D(L_2)$.

22. Let L_1 and L_2 be distributive lattices with pseudocomplementation and
let φ be a $\{0, 1\}$-homomorphism of L_1 into L_2. Prove that φ is an algebra
homomorphism iff $S(L_1)\varphi \subseteq S(L_2)$ and $D(L_1)\varphi \subseteq D(L_2)$ (H. Lakser).

16.   Representation Theorems

The Birkhoff-Stone Theorem (Theorem 7.19), which claims that every
distributive lattice can be represented by sets, is the model for the representation theorems we want to prove in this section. All such theorems are
based on a universal algebraic result, the Subdirect Representation
Theorem of G. Birkhoff [1944]. In certain special cases its use can be
avoided—we gave a direct proof for distributive lattices and we could give

a direct proof for Stone algebras. However, avoidance of it in this section is neither possible nor desirable. So we begin by developing universal algebraic tools to deal with such problems. As usual, for our purposes, "algebra" could be read "lattice" or "lattice with pseudocomplementation."

DEFINITION 1. *An algebra A is called* subdirectly irreducible *if there exist elements $u,v \in A$ such that $u \neq v$ and $u \equiv v(\Theta)$ for all congruences $\Theta > \omega$.*

In other words, $C(A) = \{\omega\} \cup [\Theta(u, v))$. An equivalent form is:

COROLLARY 2. *The algebra A is subdirectly irreducible iff $\omega = \bigwedge(\Theta_i \mid i \in I)$ ($\Theta_i \in C(A)$ for $i \in I$) implies that $\Theta_i = \omega$ for some $i \in I$.*

EXAMPLE 3. A distributive lattice L is subdirectly irreducible iff $|L| = 2$.

PROOF. If $|L| = 1$, then L is not subdirectly irreducible by definition. If $|L| = 2$, then obviously L is subdirectly irreducible. Let $|L| > 2$. Then there exist $a,b,c \in L$, $a < b < c$. We claim that $\Theta(a, b) \wedge \Theta(b, c) = \omega$, which by Corollary 2 shows that L is not subdirectly irreducible. Let $x \equiv y(\Theta(a, b) \wedge \Theta(b, c))$. By Theorem 9.3 this implies that $x \vee b = y \vee b$ and $x \wedge b = y \wedge b$; thus $x = y$ by Corollary 7.3. ●

EXAMPLE 4. $\mathfrak{B}_2$ is the only subdirectly irreducible Boolean algebra.

PROOF. Let B be Boolean. The statement is obvious for $|B| \leq 2$. If $|B| > 2$, then B has a direct product representation, $B = B_1 \times B_2$, $|B_1|$, $|B_2| \geq 2$; thus B cannot be subdirectly irreducible. ●

Equational classes of universal algebras can be introduced by defining polynomials and identities, just as in the case of lattices. However, in the next theorem the reader can avoid the use of this terminology by substituting for "equational class" the phrase "class closed under the formation of subalgebras, homomorphic images, and direct products." (This, incidentally, is an equivalent formulation by G. Birkhoff [1935]; such an equivalence will be established for the classes of algebras we consider. For the general case see exercises 6–12 and G. Grätzer [1968], Theorem 26.3.)

THEOREM 5 (G. Birkhoff [1944]).   *Let* **K** *be an equational class of algebras.*
Every algebra A in **K** *can be embedded in a direct product of subdirectly*
irreducible algebras in **K**.

PROOF.   For $a,b \in A$, $a \neq b$, let $\mathscr{X}$ denote the set of all congruences Θ of A
satisfying $a \not\equiv b(\Theta)$. Since $\omega \in \mathscr{X}$, $\mathscr{X}$ is not empty. Let C be a chain in Φ.
Then $\Theta = \bigcup (\Phi \mid \Phi \in C)$ is a congruence, $a \not\equiv b(\Theta)$, and thus every chain
in $\mathscr{X}$ has an upper bound. By Zorn's Lemma, $\mathscr{X}$ has a maximal element
$\Psi(a, b)$. We claim that $A/\Psi(a, b)$ is subdirectly irreducible; in fact, $u =$
$[a]\Psi(a, b)$ and $v = [b]\Psi(a, b)$ satisfy the condition of Definition 1.
Indeed, if Θ is any congruence of $A/\Psi(a, b)$, $\Theta \neq \omega$, then by Lemma 15.8,
$\Theta = \Phi/\Psi(a, b)$. Since $\Theta \neq \omega$, we obtain $\Phi > \Psi(a, b)$, and so $a \equiv b(\Phi)$.
Thus $u \equiv v(\Theta)$, as claimed.

   Let $B = \prod (A/\Psi(a, b) \mid a,b \in A, a \neq b)$; then B is a direct product of
subdirectly irreducible algebras. We embed A into B by $\varphi: x \to f_x$, where
f_x takes on the value $[x]\Psi(a, b)$ in the algebras $A/\Psi(a, b)$. Clearly, φ is a
homomorphism. To show that φ is one-to-one, assume that $f_x = f_y$. Then
$x \equiv y(\Psi(a, b))$ for all $a,b \in A$, $a \neq b$. Therefore, $x \equiv y(\bigwedge (\Psi(a, b) \mid a,b \in$
$A, a \neq b)$, and so $x \equiv y(\omega)$, that is, $x = y$. ●

   We got a little bit more than claimed. If we pick $x \in A/\Psi(a, b)$, then
$x = [y]\Psi(a, b)$ for some $y \in A$. Thus there is an element in the representation of A whose component in $A/\Psi(a, b)$ is x; such a representation is
called *subdirect.*

COROLLARY 6.   *In an equational class* **K**, *every algebra has a subdirect*
representation by subdirectly irreducible algebras in **K**.

   Observe how strong Theorem 5 is. If combined with Example 3 it
yields Theorem 7.19; when we combine it with Example 4 we obtain
Corollary 7.21.

   The main result of this section is the determination of subdirectly
irreducible distributive lattices with pseudocomplementation. For a
Boolean algebra B with smallest element 0 and unit e, set $\hat{B} = B \cup \{1\}$ and
let $x < 1$ for all $x \in B$. Note that if $B = (\mathfrak{B}_2)^n$, then $\hat{B}$ is the algebra $\mathfrak{S}_n$
defined in Section 15. Again, for $x \in \hat{B}$,

$$x^* = \begin{cases} 1 & \text{if } x = 0 \\ x' & \text{if } x \in B, x \neq 0 \\ 0 & \text{if } x = 1 \end{cases}$$

THEOREM 7 (H. Lakser [a]).   *A distributive lattice with pseudocomplementation L is subdirectly irreducible iff L is isomorphic to $\hat{B}$ for some Boolean algebra B.*

REMARK.   For finite L this result was proved independently by K. B. Lee [1970].

PROOF.   In $\hat{B}$ let $a = e$, let $b = 1$, and let Θ be any congruence other than ω. Then there exist $x,y \in \hat{B}$, $x < y$, such that $x \equiv y(\Theta)$. If $y = 1$, then $x \vee e \equiv y \vee e(\Theta)$, that is, $a \equiv b(\Theta)$. If $y \neq 1$, then $y \leq e$. Set $z = x' \wedge y$. Then $z \equiv 0(\Theta)$, and thus $z^* \equiv 0^*(\Theta)$; that is, $z' \equiv 1(\Theta)$, which again implies that $a \equiv b(\Theta)$.

Conversely, let L be a subdirectly irreducible distributive lattice with pseudocomplementation. Then $|L| \geq 2$ by Definition 1. Let $a,b \in L$, $a < b$ be the elements establishing that L is subdirectly irreducible and let $\langle \Theta_1, \Theta_2 \rangle$ be the congruence pair describing $\Theta(a, b)$.

If $D(L) = \{1\}$, then L is Boolean. (By Theorem 15.1(xvi) for all $x \in L$, $x \vee x^* \in D(L) = \{1\}$; that is, $x \vee x^* = 1$, and so x^* is the complement of x.) Thus $|L| = 2$ by Example 4, and so $L = \hat{B}$, where B is a one-element Boolean algebra.

Therefore, we can assume that $|D(L)| > 1$. Let Φ be a nontrivial congruence on $D(L)$. Then $\langle \omega, \Phi \rangle$ is a congruence pair that represents a nontrivial congruence, and thus we must have $\Theta_1 \leq \omega$ and $\Theta_2 \leq \Phi$. We conclude that $\Theta_1 = \omega$ and that Θ_2 is contained in every nontrivial congruence of $D(L)$. Therefore, $D(L)$ is a subdirectly irreducible distributive lattice, and so by Example 3, $|D(L)| = 2$; let $D(L) = \{e, 1\}$. We conclude that $a = e, b = 1$.

Let $x \in L$, $x \neq 1$, and $x \nleq e$. Then by Theorem 15.1(xvi), $x \vee x^* \in \{e, 1\}$. Since $x \nleq e$, we conclude that $x \vee x^* = 1$, and so $x \in S(L)$. Let Ψ be the congruence on $S(L)$ determined by $[x]$. If $y \in S(L)$, $u \in D(L)$, $y \leq u$, and $y \equiv 1(\Psi)$, then $x \leq y$, and $x \leq u \in D(L)$, so we can conclude that $u = 1$. Thus $\langle \Psi, \omega \rangle$ is a congruence pair that determines a nontrivial congruence that does not collapse e and 1, a contradiction. Therefore, for $x \in L$ we obtain $x \leq e$ for all $x \neq 1$. Since for $x \leq e$ we have $x^* \leq e$, we conclude that $x \vee x^* = e$. For $x \leq e$, define $x' = x^*$ if $x \neq 0$ and $x' = e$ if $x = 0$. Then $B = \langle (e]; \wedge, \vee, ', 0, e \rangle$ is a Boolean algebra and $L = \hat{B}$.   ●

Our next task is to describe all equational classes of distributive lattices with pseudocomplementation. We start with a few examples: $\mathbf{B}_{-1}$ is the class of one-element lattices with pseudocomplementation; let $\mathbf{B}_0$ denote the class of Boolean algebras, and for $n \geq 1$, let $\mathbf{B}_n$ be the class of all distributive lattices with pseudocomplementation satisfying the identity $(\mathbf{L}_n)$ (defined in Section 15). It follows from exercise 6.23 that there is a set of identities Σ (in terms of polynomials built up by $\wedge$, $\vee$, and $*$) such that Σ characterizes distributive lattices with pseudocomplementation (P. Ribenboim [1949]). Let $\mathbf{B}_\omega$ be the class determined by Σ. Thus $\mathbf{B}_n$ for $n \geq 1$ is defined by $\Sigma \cup \{\mathbf{L}_n\}$. $\mathbf{B}_{-1}$ is defined by $x = y$, and $\mathbf{B}_0$ is defined by $\Sigma \cup \{\mathbf{L}_0\}$, where

$(\mathbf{L}_0)$ $\qquad\qquad\qquad x \vee x^* = 1$

defines Boolean algebras.

It is obvious from Lemma 15.2 that $(\mathbf{L}_n)$ implies $(\mathbf{L}_{n+1})$ for $n \geq 1$; this implication is also obvious for $n = 0, 1$. Lemma 15.3 shows that $\mathbf{B}_n \neq \mathbf{B}_{n+1}$ for $n \geq 1$; it is obvious for $n = 0, 1$. Thus

$$\mathbf{B}_{-1} \subset \mathbf{B}_0 \subset \cdots \subset \mathbf{B}_n \subset \mathbf{B}_{n+1} \subset \cdots \subset \mathbf{B}_\omega$$

is a strictly increasing sequence of equational classes of distributive lattices with pseudocomplementation.

THEOREM 8 (K. B. Lee [1970]). *The* $\mathbf{B}_n$ $(-1 \leq n \leq \omega)$ *are the only equational classes of distributive lattices with pseudocomplementation. Moreover, if* $n \geq 0$ *and* $A \in \mathbf{B}_n$, *then* A *is a subdirect product of copies of* $\mathfrak{S}_0, \ldots, \mathfrak{S}_n$.

PROOF (H. Lakser [a]). Let $\mathbf{K}$ be a class of distributive lattices with pseudo-complementation closed under the formation of subalgebras, homomorphic images, and direct products, and let $\mathbf{S}$ denote the class of subdirectly irreducible algebras in $\mathbf{K}$. Let $\mathbf{S}_n$ denote the subdirectly irreducible algebras in $\mathbf{B}_n$. Combining Lemma 15.3 and Theorem 7, we obtain (up to isomorphism) for $n < \omega$

$$\mathbf{S}_n = \{\mathfrak{S}_0, \ldots, \mathfrak{S}_n\},$$

and $\mathbf{S}_\omega$ is the class of all subdirectly irreducible algebras.

If $\mathfrak{S}_n \in \mathbf{S}$, then $\mathbf{S}_n \subseteq \mathbf{S}$ since $\mathfrak{S}_0, \ldots, \mathfrak{S}_{n-1}$ are subalgebras of $\mathfrak{S}_n$. We have to distinguish two cases.

CASE 1.   *There is a largest integer N such that $\mathfrak{S}_N \in \mathbf{S}$. We claim that in this case $\mathbf{K} = \mathbf{B}_N$.*

CASE 2. *There is no largest integer N with $\mathfrak{S}_N \in S$. Then we claim that* $\mathbf{K} = \mathbf{B}_\omega$.

PROOF OF CASE 1. By assumption, $S_N \subseteq S$ and $\mathfrak{S}_{N+1} \notin S$. If $A \in S - S_N$, then $A = \hat{B}$ for some Boolean lattice B with $|B| > 2^N$; thus $\mathfrak{S}_{N+1}$ is a subalgebra of A,$\mathfrak{S}_{N+1} \in S$, a contradiction. Therefore, $S = S_N$ and so, by Theorem 5, $\mathbf{K} = \mathbf{B}_N$. ▶

PROOF OF CASE 2. In this case $\hat{B} \in S$ for any finite Boolean lattice B. To show that $S = S_\omega$ and therefore that $\mathbf{K} = \mathbf{B}_\omega$, it suffices to show that if B is any Boolean lattice, then $\hat{B} \in S$. Indeed, let $\mathscr{X}$ denote the family of finite subalgebras of B; for $B_1 \in \mathscr{X}$, let $\hat{B}_1$ denote $B_1 \cup \{1\}$, where 1 is the unit element of $\hat{B}$. Then $\hat{B}_1$ is a subalgebra of $\hat{B}$. Set $\hat{\mathscr{X}} = \{\hat{B}_1 \mid B_1 \in \mathscr{X}\}$. Observe that each $A \in \hat{\mathscr{X}}$ is finite; thus $\hat{\mathscr{X}} \subseteq S$, $\hat{\mathscr{X}}$ is directed, and $\bigcup (A \mid A \in \mathscr{X}) = \hat{B}$. Since these imply that $\hat{B} \in K$ (see exercises 8–12), we conclude that $\hat{B} \in S$. ●

The representation theorem of Section 7 stated that every Boolean algebra can be embedded in some $P(X)$. Verifying a conjecture of O. Frink [1962], G. Grätzer [1963] proved that every Stone algebra can be embedded in some $I(P(X))$, viewed as a distributive lattice with pseudocomplementation. This cannot be improved upon because, according to exercise 14.5, $I(P(X))$ is a Stone algebra. It was conjectured in G. Grätzer [1963] that by replacing $P(X)$ by a (noncomplete) atomic Boolean algebra, the previous result extends to all distributive lattices with pseudocomplementation. This idea was proved in H. Lakser [a], utilizing Theorem 7:

THEOREM 9. *Let L be a distributive lattice with pseudocomplementation. Then there is an atomic Boolean algebra B such that L can be embedded into $I(B)$.*

PROOF. Let L be subdirectly irreducible. Then $L = \hat{B}_0$ for some Boolean lattice B_0. By Corollary 7.21 there is an embedding $\varphi: B_0 \to P(X)$. We thus obtain an embedding $\hat{\varphi}: L \to \widehat{P(X)}$ by setting $a\hat{\varphi} = a\varphi$ for $a \in B_0$ and $1\hat{\varphi} = 1$. Choose an infinite set Y and let B be the Boolean algebra of all finite and cofinite (that is, complements of finite) subsets of $X \times Y$. Obviously, B is atomic. For $A \subseteq X$ define

$$A\psi = \{Z \mid Z \in B, Z \subseteq A \times Y, Z \text{ finite}\};$$

$$1\psi = B.$$

Then ψ maps $\widehat{P(X)}$ into $I(B)$. Observe that $\varnothing\psi = \varnothing$ and

$$X\psi = \{Z \mid Z \in B, Z \text{ finite}\}.$$

The formulas $(X \cap Y)\psi = X\psi \wedge Y\psi$ and $(X \cup Y)\psi = X\psi \vee Y\psi$ are immediate. Thus ψ is a lattice-embedding of $\widehat{P(X)}$ into $I(B)$. Let $A \subseteq X$; then

$$(A\psi)^* = \{Z \mid Z \in B, Z \cap W = \varnothing \text{ for all } W \in A\psi\}$$
$$= \{Z \mid Z \in B, Z \subseteq (X - A) \times Y\}.$$

Thus if $A = \varnothing$, then $(A\psi)^* = B$. If $A \neq \varnothing$, then the complement of $(X - A) \times Y$ in $X \times Y$ is infinite (Y is infinite); therefore, $Z \subseteq (X - A) \times Y$ with $Z \in B$ cannot be cofinite, so such a Z must be finite. Consequently,

$$(A\psi^*) = \{Z \mid Z \subseteq (X - A) \times Y, \ Z \text{ finite}\} = (X - A)\psi.$$

Thus ψ is an embedding of the algebra $\widehat{P(X)}$ into $I(B)$, and $\phi\psi$ embeds L into $I(B)$.

Since the existence of an embedding into some $I(B)$ is preserved under the formation of subalgebras (which is obvious) and direct products (see exercise 16), a reference to Theorems 5 and 7 concludes the proof. ●

Exercises

1. Show that every *simple algebra* A (that is, $|C(A)| = 2$) is subdirectly irreducible.

2. Show that for every cardinal $\mathfrak{m} > 4$, there is a simple lattice of cardinality $\mathfrak{m}$.

3. Verify the statement of Example 3 without any reference to Section 9.

*4. Show that Theorem 5 is equivalent to the Axiom of Choice (G. Grätzer, *Notices Amer. Math. Soc.* 14 (1967), 133).

5. Let B be a Boolean algebra. Show that every subalgebra of a $\hat{B}$ is either Boolean or of the form $\hat{B}_1$, where B_1 is a subalgebra of B.

In exercises 6–12, let K be a class of algebras closed under the formation of subalgebras, homomorphic images, and direct products.

6. Let A be an algebra and let $\mathscr{A}$ be a family of subalgebras of A such that for $B \in \mathscr{A}$ we have $B \in \mathbf{K}$, $\bigcup (X \mid X \in \mathscr{A}) = A$, and for $X, Y \in \mathscr{A}$ there

is a $Z \in \mathscr{A}$ such that $X \subseteq Z$ and $Y \subseteq Z$. Define $B \subseteq \prod (X \mid X \in \mathscr{A})$ by $f \in B$ iff there exists an $X \in \mathscr{A}$ such that for all $Y, Z \in \mathscr{A}$ with $X \subseteq Y$ and $X \subseteq Z$, we have $f(Y) = f(Z)$. Show that B is a subalgebra of $\prod (X \mid X \in \mathscr{A})$, and therefore $B \in \mathbf{K}$.

7. Define a relation Θ on the algebra B of exercise 6: $f \equiv g(\Theta)$ iff there exists an $X \in \mathscr{A}$ such that for all $Y \in \mathscr{A}$ with $Y \supseteq X$, we have $f(Y) = g(Y)$. Show that Θ is a congruence relation on B and that $A \cong B/\Theta$. Conclude that $A \in \mathbf{K}$.

8. Let $F_{\mathbf{K}}(\omega)$ denote the free algebra over $\mathbf{K}$ freely generated by $x_0, x_1, \ldots,$ $x_n, \ldots$. Let Σ denote the set of all identities $p = q$ such that

$$p(x_0, x_1, \ldots) = q(x_0, x_1, \ldots).$$

Show that the identities in Σ hold in $F_{\mathbf{K}}(\omega)$.

9. For an ordinal α let $F_{\mathbf{K}}(\alpha)$ denote the free algebra over $\mathbf{K}$ freely generated by $x_0, \ldots, x_\gamma, \ldots, \gamma < \alpha$. Show that Σ holds in $F_{\mathbf{K}}(\alpha)$.

10. Show that every $A \in \mathbf{K}$ is a homomorphic image of some $F_{\mathbf{K}}(\alpha)$. Conclude that Σ is satisfied in every $A \in \mathbf{K}$.

11. Show that every algebra A satisfying Σ is the homomorphic image of some $F_{\mathbf{K}}(\alpha)$.

12. Combine exercises 8–11 to show that $\mathbf{K}$ is an equational class of algebras. (This is G. Birkhoff's characterization theorem for equational classes; see G. Birkhoff [1935].)

13. Show that a distributive lattice with pseudocomplementation L belongs to $\mathbf{B}_n$ ($1 \leq n < \omega$) iff every prime ideal of L contains at most n distinct prime ideals (K. B. Lee [1970]).

14. A congruence relation Θ is *completely meet-irreducible* if $\Theta = \bigwedge (\Theta_i \mid i \in I)$ implies that $\Theta = \Theta_i$ for some $i \in I$. Show that Theorem 5 is equivalent to the statement that for any algebra A, any congruence is the complete meet of completely meet-irreducible congruences.

15. Combine exercises 13 and 14 to show the Congruence Extension Property for $\mathbf{B}_n$ ($1 \leq n < \omega$).

16. Let the L_i, $i \in J$, be distributive lattices with 0. Define the mapping $\varphi \colon \prod (I(L_i) \mid i \in J) \to I(\prod (L_i \mid i \in J))$ as follows: If $f \in \prod (I(L_i) \mid i \in J)$, then $f(i)$ is an ideal of L_i for each $i \in J$; set $f\varphi = \prod (f(i) \mid i \in J)$. Show that φ embeds the distributive lattice with pseudocomplementation $\prod (I(L_i) \mid i \in J)$ into $I(\prod (L_i \mid i \in J))$.

17. Let X be an infinite set. Define $\psi \colon \mathfrak{S}_1 \to I(P(X))$ by $0\psi = \{\varnothing\}$, $1\psi = P(X)$, $e\psi = \{Z \mid Z \subseteq X, Z \text{ finite}\}$. Show that ψ embeds the algebra $\mathfrak{S}_1$ into $I(P(X))$.

18. Show that an algebra $\langle L; \wedge, \vee, *, 0, 1 \rangle$ is a Stone algebra iff it can be embedded in some $I(B)$ where B is an atomic and complete Boolean lattice (G. Grätzer [1963]).

19. Show that the free algebra over $\mathbf{B}_n$ with m generators is finite, provided that $n, m < \omega$.

20. Does the statement of exercise 19 hold for $\mathbf{B}_\omega$?
21. Show that a finitely generated subalgebra of a distributive lattice with pseudocomplementation is finite (K. B. Lee [1970]).
22. Find a direct proof of the statement that every Stone algebra is a subdirect product of copies of $\mathfrak{S}_0$ and $\mathfrak{S}_1$ (use exercise 15.11).

17. Injective and Free Stone Algebras

Many of the categorical concepts introduced in Section 13 can be fruitfully studied in the categories $\mathbf{B}_n$ for $0 \le n \le \omega$. To illustrate this we shall investigate injectives in $\mathbf{B}_1$, proving the Characterization Theorem of R. Balbes and G. Grätzer [a]. Further developments along this and other lines are mentioned in the exercises and in the Further Topics and References at the end of this section.

Let us recall that a Stone algebra I is *injective* if, whenever A and B are Stone algebras and B is a subalgebra of A, then any homomorphism β of B into I can be extended to a homomorphism α of A into I.

We shall need two simple facts about injective algebras.

LEMMA 1.

(i) *A direct product of injective algebras is injective.*

(ii) *A retract of an injective algebra is injective.*

PROOF. Let A and B be Stone algebras and let B be a subalgebra of A.

(i) Let $I_j, j \in J$, be injective algebras and let β be a homomorphism of B into $\prod (I_j \mid j \in J)$. Define $\beta_j: B \to I_j$ by $x\beta_j = (x\beta)(j)$, that is, β followed by the jth projection. Since I_j is injective, there exists an extension α_j of β_j to A. Define $\alpha: A \to \prod (I_j \mid j \in J)$ by $(x\alpha)(j) = x\alpha_j$ for $j \in J$. Then α is a homomorphism extending β to A. ◗

(ii) Let I be an injective algebra, let ρ be a retraction on I, and let β be a homomorphism of B into $I\rho$. We can view β as a homomorphism of B into I; thus β has an extension α_1 to A. It is easily seen that $\alpha = \alpha_1\rho$ is an extension of β to A and that it maps A into $I\rho$. ●

Our results on injective Stone algebras can be summarized in the form of two theorems, the first of which can be called the Characterization Theorem, the second of which can be called the Representation Theorem (both are from R. Balbes and G. Grätzer [a]):

THEOREM 2. *A Stone algebra I is injective iff it satisfies the following conditions:*

   (i) *I is complete.*

   (ii) *I has a smallest dense element d.*

   (iii) *Every element a has a dual pseudocomplement a^+ (that is, $a \vee x = 1$ iff $x \geq a^+$) and $a^+ \wedge a^{++} = 0$ (that is, $a^+ \in S(I)$).*

   (iv) *$a^* = b^*$ and $a^+ = b^+$ imply that $a = b$.*

Let us recall a construction of Stone algebras (exercise 14.10): For a Boolean algebra B let $B^{[2]} = \{\langle a, b \rangle \mid a \leq b, a,b \in B\}$; observe that $B^{[2]}$ is a sublattice of B^2 and that $\langle a, b \rangle^* = \langle b', b' \rangle$; thus $\langle a, b \rangle^* \vee \langle a, b \rangle^{**} = \langle b', b' \rangle \vee \langle b, b \rangle = \langle 1, 1 \rangle$. Therefore, $B^{[2]}$ is a Stone algebra.

THEOREM 3. *A Stone algebra I is injective iff it can be represented in the form $B_0 \times B_1^{[2]}$, where B_0 and B_1 are complete Boolean algebras. In this representation B_0 and B_1 are uniquely determined up to isomorphism.*

We prepare the proofs of Theorems 2 and 3 in a series of ten steps. The proofs of the theorems will be easy combinations of the following statements.

   (a) A prime ideal of a Stone algebra contains one and only one minimal prime ideal. This is an obvious combination of Lemma 15.6 and the case $n = 1$ of Theorem 16.8. ▶

   (b) $\mathfrak{S}_1$ is injective. Let A and B be Stone algebras, let B be a subalgebra of A, and let β be a homomorphism of B into $\mathfrak{S}_1$. Set $Q = \{x \mid x \in B, x\beta = 0\}$ and $P = \{x \mid x\beta \leq e\}$. By Lemma 15.6, P and Q are prime ideals and Q is the minimal prime ideal contained in P. Again by Lemma 15.6, it suffices to find prime ideals P_1 and Q_1 of A, Q_1 minimal in P_1, such that $P = P_1 \cap B$ and $Q = Q_1 \cap B$. By Theorem 7.15 there exists a prime ideal P_1 of A containing $(P]_A$ and disjoint to $[B - P)_A$. Then $P_1 \cap B = P$. Let Q_1 denote the minimal prime ideal of A contained in P_1. Then $Q_1 \cap B$ is minimal in B by Lemma 15.5, and $Q_1 \cap B \subseteq P_1$; thus by (a), $Q_1 \cap B = Q$. ▶

   (c) $\mathfrak{S}_0$ is injective. This is obvious by (b) and by Lemma 1(ii). ▶

   (d) A Stone algebra I is injective iff it is a retract of some $\mathfrak{C}_0^m \times \mathfrak{C}_1^n$. The "if" part follows from (b), (c), Lemma 1(i) and (ii). The "only if" part follows from Theorem 16.8 and the fact that an injective algebra is a retract of any extension. ▶

(e) An injective algebra I satisfies (i)–(iii) of Theorem 2, and furthermore, $D(I)$ is Boolean. Indeed, $\mathfrak{C}_0^m \times \mathfrak{C}_1^n$ satisfies (i)–(iii) of Theorem 2 and the dense set is Boolean; these properties are preserved under retraction, and thus (d) completes the proof. ◗

(f) An injective Stone algebra I satisfies (iv) of Theorem 2. Let $a,b \in I$, $a \neq b$, $a^* = b^*$, and $a^+ = b^+$. Take a prime ideal P containing exactly one of a and b—for example, $a \in P$, $b \notin P$. Then $a^{**} = b^{**} \geq b \notin P$, and thus $a^{**} \notin P$. By Lemma 15.5, P is not minimal. Let Q be the minimal prime ideal in P. Using the dual argument (justified by condition (iii) of Theorem 2 as proved in (e)), there exists a prime ideal R such that $R \supset P$ and $L - R$ is a minimal prime dual ideal, that is, R is a maximal prime ideal. Pick an $x \in P \cap D(I)$ (see Lemma 15.5) and a $y \in R - P$ with $x < y$. Consider the map $\varphi: a \to 0$ for $a \in Q$; $a \to x$ for $a \in P - Q$; $a \to y$ for $a \in R - P$; $a \to 1$ for $a \in L - R$. It is easily seen (just as in the proof of Lemma 15.6) that φ is a retraction; therefore, by Lemma 1(ii), the four-element chain is injective, contradicting the last statement of (e). (The dense set is the three-element chain, which is not Boolean.) ◗

(g) Let A satisfy (i)–(iv) of Theorem 2; then A can be represented in the form $B_0 \times B_1^{[2]}$, where B_0 and B_1 are complete Boolean algebras. Indeed, $A \cong (a] \times (a^*]$, where $a = d^{++}$ (d is given by (ii)). Since by (iii) d^{++} has a complement (namely, d^+), we obtain $a \in S(A)$, hence the possibility of such a decomposition. The pseudocomplement of $b \in (A]$ is $b^0 = a \wedge b^*$, so $b \vee b^0 = b \vee (a \wedge b^*) = (b \vee b^*) \wedge (b \vee a) \geq a$, since $b \vee b^* \in D(A)$, and so $b \vee b^* \geq d \geq d^{++} = a$. On the other hand, $b \vee b^0 \leq a$, and thus $b \vee b^0 = a$, verifying that $(a]$ is a Boolean algebra. By (i), $(a]$ is complete; so is $(a^*]$. Therefore, it suffices to show that $(a^*] \cong (S((a^*]))^{[2]}$. For $x \in (a^*]$, let x^0 denote the pseudocomplement of x in $(a^*]$, that is, $x^0 = a^* \wedge x^*$. We claim that the dual pseudocomplement x^∞ of x in $(a^*]$ is $x^\infty = x^+ \wedge a^*$. Indeed, $x \vee x^\infty = x \vee (x^+ \wedge a^*) = (x \vee x^+) \wedge (x \vee a^*) = 1 \wedge a^* = a^*$; if $x \vee y = a^*$, then $x \vee y \vee a^{**} = a^* \vee a^{**} = 1$, and so $y \geq (x \vee a^{**})^+ = x^+ \wedge (a^{**})^+ = x^+ \wedge a^* = x^\infty$. The map $x \to \langle x^{\infty\infty}, x^{00} \rangle$ embeds $(a^*]$ into $(S((a^*]))^{[2]}$ by (iii) and (iv). The map is onto because if $\langle b, c \rangle \in (S((a^*]))^{[2]}$, that is, $b,c \in S((a^*])$, $b \leq c$, then $x \to \langle b, c \rangle$, where $x = (b \vee d) \wedge c$. Indeed, using $d^{++} = a$ and $d^* = 0$, we obtain $x^{\infty\infty} = (x^+ \wedge a^*)^\infty = (x^+ \wedge a^*)^+ \wedge a^* = (x^{++} \vee a^{**}) \wedge a^* = (((b^{++} \vee d^{++}) \wedge c^{++}) \vee a^{**}) \wedge a^* = (((b \vee a) \wedge c) \vee a) \wedge a^* = (b \vee a) \wedge a^* = (b \wedge a^*) \vee (a \wedge a^*) = b \vee 0 = b$; similarly, $x^{00} = c$. ◗

(h) Every complete Boolean algebra is injective. Every complete

Boolean algebra is a retract of some power of $\mathfrak{S}_0$ by Corollary 7.21 and Theorem 13.14; thus Lemma 1(i) and (ii) complete the proof. ▶

(j) If B is a complete Boolean algebra, then $B^{[2]}$ is injective. By Theorem 16.8, $B^{[2]}$ can be embedded in some $A = \mathfrak{C}_0^m \times \mathfrak{C}_1^n$. Thus $B \cong S(B^{[2]})$ is a subalgebra of $S(A)$, and so, by Theorem 13.14, there is a retraction $\psi: S(A) \to S(B^{[2]})$. Then for $a \in A$ define $a\varphi = \langle a^{++}\psi, a^{**}\psi \rangle$. It is easily seen that φ is a retraction of A onto $B^{[2]}$; therefore, $B^{[2]}$ is injective by (d). ▶

(k) If B_0 and B_1 are complete Boolean algebras, then $B_0 \times B_1^{[2]}$ is injective. This is proved trivially, by (h), (j), and Lemma 1(i). ▶

PROOF OF THEOREM 2.   If I is injective, then I satisfies (i)–(iv) of Theorem 2 by (e) and (f). Conversely, if I satisfies (i)–(iv), then by (g), $I \cong B_0 \times B_1^{[2]}$, where B_0 and B_1 are complete Boolean algebras, and so I is injective by (k). ●

PROOF OF THEOREM 3.   The first statement has already been proved in verifying Theorem 2. The uniqueness of the representation follows from the fact that $B_0 \times B_1^{[2]}$ has a smallest dense element d, and B_0 is (lattice) isomorphic to $(d^{++}]$ and B_1 is (lattice) isomorphic to $[d)$. ●

The second topic to be taken up is free products and free algebras. Although we prove the existence of free products in all classes $\mathbf{B}_n$, $0 \leq n \leq \omega$, we shall investigate it in detail only for Stone algebras.

LEMMA 4.   *Let A_i, $i \in I$, be algebras of more than one element in $\mathbf{B}_n$, where $0 < n \leq \omega$. Then a free $\mathbf{B}_n$-product of A_i, $i \in I$, exists.*

PROOF.   By the proof of Theorem 13.18 or, more explicitly, by exercise 12.9, we have only to prove that if $a,b \in A_k$ and $a \neq b$, then for some $A \in \mathbf{B}_n$ there exist homomorphisms $\varphi_i: A_i \to A$ for all $i \in I$ such that $a\varphi_k \neq b\varphi_k$. By Theorems 16.5 and 16.7, there exists a Boolean algebra B such that there is a homomorphism $\varphi: A_k \to \hat{B}$ satisfying $a\varphi \neq b\varphi$. Set $\varphi_k = \varphi$. By Lemmas 15.4 and 15.6 (case $n = 1$, $P = Q_1$), every algebra A_i has a homomorphism φ_i into $\hat{B}$ (in fact, $A_i\varphi_i = \{0, 1\}$) so choose this φ_i for $i \neq k$. ●

For Stone algebras the situation is surprisingly simple (G. Grätzer and H. Lakser [1969b]):

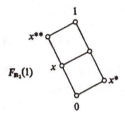

Figure 17.1

THEOREM 5.   *Let A_i, $i \in I$, be Stone lattices. Regarding them as distributive lattices with 0 and 1, form a free $\{0, 1\}$-distributive product A of the A_i, $i \in I$. Then A is a Stone lattice. Furthermore, A as a Stone algebra is a free (Stone) product of A_i, $i \in I$.*

PROOF.   Let L be a distributive lattice with 0 and 1 and let $\mathscr{S}(L)$ be the Stone space of L. An ideal I of L is represented by an open set $r(I)$ of $\mathscr{S}(L)$. For $a \in L$, $r((a]^*)$ is the *interior* of $\mathscr{S}(L) - r(a)$ (that is, the largest open set contained in $\mathscr{S}(L) - r(a)$). Since L is Stone iff $(a]^* = (b]$ for some b that is complemented, this translates into the condition that $\mathscr{S}(L) - r(a)$ be clopen, or equivalently:

(S4)   The closure of a compact open set is open.

Clearly, (S4), as well as (S1)–(S3) of Section 11, are preserved under topological product. Thus A is a Stone lattice.

   To show that A is a free Stone product, let D be a Stone algebra, and for each $i \in I$ let φ_i be a homomorphism of the Stone algebra A_i into D. Since A is a free $\{0, 1\}$-distributive product of the A_i, $i \in I$, there exists a $\{0, 1\}$-homomorphism φ extending all the φ_i, $i \in I$. The fact that $(x\varphi)^* = x^*\varphi$ remains to be proven. We know this to hold for $x \in A_i$. Let $x \in A$, $x = \bigvee \bigwedge x_{\alpha\beta}$, where $x_{\alpha\beta} \in \bigcup (A_i \mid i \in I)$. Then

$$(x\varphi)^* = ((\bigvee \bigwedge x_{\alpha\beta})\varphi)^* = (\bigvee \bigwedge x_{\alpha\beta}\varphi)^*$$

$$= \bigvee \bigwedge (x_{\alpha\beta}\varphi)^* = \bigvee \bigwedge x_{\alpha\beta}^*\varphi = (\bigvee \bigwedge x_{\alpha\beta}^*)\varphi = x^*\varphi. \quad \blacksquare$$

   The free Stone algebra on one generator $F_{\mathbf{B}_1}(1)$ is shown in Figure 17.1. Observe that $F_{\mathbf{B}_1}(1) = \mathfrak{S}_0 \times \mathfrak{S}_1$, and so $\mathscr{S}(F_{\mathbf{B}_1}(1)) = \mathscr{S}(\mathfrak{S}_0) \cup \mathscr{S}(\mathfrak{S}_1)$, where the $\cup$ is disjoint union. Since $F_{\mathbf{B}_1}(n)$ is the free product of n copies of $F_{\mathbf{B}_1}(1)$, $\mathscr{S}(F_{\mathbf{B}_1}(n))$ is $(\mathscr{S}(\mathfrak{S}_0) \cup \mathscr{S}(\mathfrak{S}_1))^n$. Note that $\mathscr{S}(\mathfrak{S}_0)$ is a singleton, and $\mathfrak{S}_1$ is the free distributive lattice with 0 and 1. Therefore, $(\mathscr{S}(\mathfrak{S}_1))^k$ is the Stone space of $F_{\mathbf{D}_{\{0,1\}}}(k)$. Thus,

THEOREM 6.   *The following isomorphism holds:*

$$F_{\mathbf{B}_1}(n) \cong \prod (F_{\mathbf{D}_{\{0,1\}}}(|X|) \mid X \subseteq \{1, \ldots, n\}).$$

This result is due to R. Balbes and A. Horn [1970a]; the present proof is from G. Grätzer and H. Lakser [1969b].

Exercises

1. Let A be a Stone lattice, let B be a Boolean lattice, and let $\varphi: A \to B$ be a $\{0, 1\}$-homomorphism. Show that $\bar{\varphi}: x \to \langle x\varphi, x^{**}\varphi \rangle$ is a homomorphism of the Stone algebra A into the Stone algebra $B^{[2]}$.

2. Show that $\varphi \to \bar{\varphi}$ establishes a one-to-one correspondence between the $\{0, 1\}$-homomorphisms of A into B and the algebra homomorphisms of A into $B^{[2]}$.

3. Use exercises 1 and 2 to show that, for a complete Boolean lattice B, the algebra $B^{[2]}$ is injective. (This proof is due to H. Lakser [1970]).

4. For a Boolean lattice B and integer $n \geq 2$, define

$$B^{[n]} = \{\langle x, y_1, \ldots, y_{n-1} \rangle \mid x, y_1, \ldots, y_{n-1} \in B, x \leq y_1, \ldots, x \leq y_{n-1}\}.$$

Partially order $B^{[n]}$ componentwise. Show that $B^{[n]}$ is a pseudocomplemented distributive lattice.

5. Show that $B^{[n]}$ as a distributive lattice with pseudocomplementation is in $\mathbf{B}_{n-1}$ and that if $|B| > 1$, then $B^{[n]} \notin \mathbf{B}_{n-2}$ (G. Grätzer and H. Lakser [b]).

6. Let $\mathbf{K}$ be a category of algebras. Let A and B be algebras in $\mathbf{K}$ and let A be a subalgebra of B. Then B is called an *essential extension of A* if, for any algebra $C \in \mathbf{K}$ and homomorphism $\varphi: B \to C$, if the restriction of φ to A is one-to-one, then φ is one-to-one. Assume that $\mathbf{K}$ is closed under the formation of homomorphic images. Show that an essential extension of a subdirectly irreducible algebra is itself subdirectly irreducible (see, for example, B. Banaschewski [1970]).

7. An injective essential extension is called an *injective hull*. Show that if B_1 and B_2 are injective hulls of A, then there is an isomorphism $\varphi: B_1 \to B_2$ that is the identity map on A.

8. Let $\mathbf{K}$ be a category of algebras closed under the formation of homomorphic images, $A, B \in \mathbf{K}$. Let A be a subalgebra of B. Show that there exists a congruence relation Θ maximal with respect to the property that $\Theta_A = \omega$. Show also that B/Θ is an essential extension (up to isomorphism) of A.

9. Let $\mathbf{K}$ be an equational class of algebras in which every algebra can be

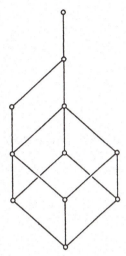

Figure 17.2

embedded in an injective algebra. Show that every algebra in **K** has an injective hull (B. Banaschewski [1970]).

*10. Determine the injective hulls of Stone algebras (H. Lakser [1970]).

*11. Verify the Amalgamation Property for Stone algebras.

*12. Verify the Amalgamation Property for $\mathbf{B}_2$.

*13. Show that the Amalgamation Property holds for $\mathbf{B}_n$ iff $n \leq 2$ or $n = \omega$ (G. Grätzer and H. Lakser [a]).

14. Show that every algebra in $\mathbf{B}_n$ can be embedded in an injective algebra iff $n \leq 2$ (A. Day [1970]).

15. An *implicational class* **K** is a class of algebras closed under isomorphism, subalgebras, and direct products. Find an implicational class **K** of distributive lattices with pseudocomplementation such that $\mathbf{K} \neq \mathbf{B}_n$ for any $n \leq \omega$ (G. Grätzer and H. Lakser [c]). (Hint: Take the smallest implicational class containing the algebra shown in Figure 17.2. Then $\mathbf{B}_2 \subset \mathbf{K} \subset \mathbf{B}_3$.)

16. Let A, B, and C be distributive lattices with pseudocomplementation, let C be a subalgebra of B, and let A be a homomorphic image of C. Prove the existence of a homomorphic image D of B such that A is a subalgebra of D.

17. Let **K** be a nontrivial equational class of lattices in which every lattice can be embedded in an injective lattice. Show that $\mathbf{K} = \mathbf{D}$ (A. Day [a]). (Hint: Use exercise 15.15.)

18. Describe $F_{\mathbf{B}_\omega}(1)$. (It has seven elements.)

19. Show that Theorem 5 does not generalize to $\mathbf{B}_2$.

20. Show that (S4) is characteristic of the Stone spaces of Stone lattices.

21. Characterize the Stone spaces of $L \in \mathbf{B}_n$.

*22. Show that $F_{\mathbf{B}_\omega}(n) \cong F_{\mathbf{B}_{2^n}}(n)$. Apply this to get a new proof of exercise 16.21 (G. Grätzer and H. Lakser).

Exercises 23–26 are from R. Balbes and G. Grätzer [a].

23. Let A and B be Stone algebras. Show that $A \times B$ is projective iff both A and B are projective.

24. Let A be a projective Stone algebra satisfying $S(A) = \{0, 1\}$. Show that $A_0 = A - \{0, 1\}$ is a sublattice of A and that A_0 is projective in $\mathbf{D}$.

25. Prove the converse of exercise 24.

26. Characterize the finite projective Stone algebras.

27. Let A be a finite Stone algebra. Why is it not true that A is projective iff $S(A)$ is projective in $\mathbf{B}$ and $D(A)$ is projective in $\mathbf{D}$?

Further Topics and References

Except for V. Glivenko's early work [1929], the study of pseudocomplemented distributive lattices started only in 1956 with a solution to Problem 70 of G. Birkhoff [1948], which gave a characterization of Stone lattices by minimal prime ideals (G. Grätzer and E. T. Schmidt [1957b]; for a simplified proof see J. Varlet [1966]). The idea of a triple was conceived by the author in 1961 as a method of proving O. Frink's conjecture on the representation of Stone lattices (see exercises 16.18). His attempt to prove the conjecture failed, however, and as a result triples were not utilized until 1968 (C. C. Chen and G. Grätzer [1969b] and [1969c]). In the sixties the structure of Stone lattices was the subject of extensive research. The following list—which is far from complete—may prove of interest to the reader: G. Bruns [1965] (simplifying the Representation Theorem of G. Grätzer [1963]—an even simpler proof is in G. Grätzer [1969]), R. Balbes and G. Grätzer [a] (characterizing finite projective Stone algebras), R. Balbes and A. Horn [1970a], O. Frink [1962], T. Katriňák [1967], H. Lakser [1970], T. Speed [1969], and J. Varlet [1963], [1966].

Originally, Chapter 3 of the present text gave a survey of this development. However, the discovery of the classes $\mathbf{B}_n$ (K. B. Lee [1970]) and the subsequent realization that many of the known facts about Stone lattices generalize to some or all the $\mathbf{B}_n$ made it necessary to revise this chapter. H. Lakser [a] generalized the Representation Theorem of G. Grätzer [1963]; the description of injectives for Stone algebras is generalized to $\mathbf{B}_2$ in G. Grätzer and H. Lakser [b]. There are no non-Boolean injectives in $\mathbf{B}_n$ for $n > 2$ (A. Day [1970]), but the description generalizes to give all absolute retracts for $n > 2$ (which is the same as injective if every algebra can be embedded in an injective). We are, of course, only at the very beginning of this development.

For a class of algebras **K**, the concept of the *Amalgamation Class of* **K**, in notation Amal (**K**), can be introduced: For $A \in$ **K** we have $A \in$ Amal (**K**) iff, for all $B,C \in$ **K**, with A a subalgebra of B and C, the amalgamation of A, B, and C can be effected in **K** in the sense of Definition 13.17. Of course, the Amalgamation Property means that Amal (**K**) = **K**. Since the Amalgamation Property fails for B_n, $2 < n < \omega$, Amal (B_n) is investigated in G. Grätzer and H. Lakser [a] to determine the extent to which B_n fails to satisfy the Amalgamation Property.

Relatively pseudocomplemented distributive lattices (often called *Heyting algebras*) arise from nonclassical logic and were first investigated by T. Skolem about 1920. For a detailed development, see H. B. Curry [1963], which contains all the important rules of computation with $a * b$. Unfortunately, the triple method is not applicable here. Recall that the effectiveness of the triple method is due to the fact that the class of Boolean algebras and the class of distributive lattices with unit are both classes with structures simpler than that of the class of distributive lattices with pseudocomplementation. Unfortunately, if L is relatively pseudocomplemented, so is $D(L)$, and consequently there is no reduction. Using equivalence classes of ordered pairs, $\langle a, b \rangle$, $a \in D(L)$, $b \in S(L)$, W. C. Nemitz [1965] developed a Construction Theorem for relatively pseudocomplemented distributive lattices. It should also be mentioned that injectives and projectives were investigated in the work of R. Balbes and A. Horn [1970b] and A. Day [1970]; see also A. Horn [1969a] and [1969b].

Pseudocomplemented distributive semilattices were investigated by T. Katriňák [1970]; unfortunately, the Construction Theorem uses equivalence classes of ordered pairs $\langle a, b \rangle$, $a \in D(L)$, $b \in S(L)$.

In connection with nonclassical logic, many algebras emerged that are distributive lattices endowed with some additional structure. The best known of these, the Post algebras, happen to be Stone lattices. In fact, a Post algebra is a free {0, 1}-distributive product of a finite chain and a Boolean algebra, in which the elements of the chain are regarded as nullary operations; see E. L. Post [1921], P. C. Rosenbloom [1942], G. Epstein [1960], T. Traczyk [1963] and [1967], C. C. Chang and A. Horn [1961], R. Balbes and Ph. Dwinger [a] and [b], and M. Mandelker [1970].

For some other types of distributive lattices with additional operations, see N. D. Belnap and J. H. Spencer [1966], J. M. Dunn and N. D. Belnap [1968], H. Rasiowa and R. Sikorski [1963], and J. Varlet [1968].

PROBLEMS

53. Describe the projective Stone algebras.
54. Describe the triples associated with free Stone algebras.
55. Find a direct (less-computational) proof of the Construction Theorem for Stone Algebras.
56. Determine the free algebras $F_{\mathbf{B}_\mathbf{k}}(n)$.
57. Let B be a Boolean lattice, let D be a distributive lattice with unit, and let A be a sublattice of $C(B) \times C(D)$. Under what conditions does there exist a distributive lattice with pseudocomplementation L such that $S(L) = B$, $D(L) = D$, and A consists of all congruence pairs of L?
58. Characterize free $\mathbf{B}_n$-products ($n > 1$).
59. Describe Amal ($\mathbf{B}_n$) for $n > 2$.
60. Determine the projectives in $\mathbf{B}_n$, $n > 1$.
61. Let $n \geq 0$. Determine all *injective structures* in the sense of J. M. Maranda [1964] in the category $\mathbf{B}_n$.
62. For every identity I for distributive lattices with pseudocomplementation, there exists a first-order sentence $\Phi(I)$ such that I holds for L iff $\Phi(I)$ holds for $\mathscr{P}(L)$. (In K. B. Lee [1970], the sentence for ($\mathbf{L}_n$) is given as follows: "Every element contains at most n minimal elements." For $n = 0$ use Theorem 7.22.) Is there a natural class of first-order sentences properly containing all identities for which the same statement can be made?
63. Show that there are $2^{\aleph_0}$ implicational classes of distributive lattices with pseudocomplementation.
64. Investigate the lattice of implicational classes of distributive lattices with pseudocomplementation.
65. Show that there are $2^{\aleph_0}$ equational classes of lattices with pseudo-complementation.
66. For a distributive lattice L, define the number $n(L)$ as follows: Let $n(L)$ be the smallest integer n such that $I(L) \in \mathbf{B}_n$ if L has a zero and $I_0(L) \in \mathbf{B}_n$ if L does not have a zero. Classify and investigate distributive lattices according to the value of $n(L)$. (The case $n = 1$ was considered by M. Mandelker [1970].)
67. Let K be an equational class of algebras satisfying $\mathbf{HS}(\mathbf{K}_1) = \mathbf{ISH}(\mathbf{K}_1)$ for all $\mathbf{K}_1 \subseteq \mathbf{K}$. Does K satisfy the Congruence Extension Property? Does the assumption that the congruence lattices of algebras in K are distributive make any difference (see exercise 15.18)?

BIBLIOGRAPHY

S. Abian and A. B. Brown
  1961. A theorem on partially ordered sets, with applications to fixed point
    theorems. *Canad. J. Math.* 13:78–82.
M. Aigner
  1969. Graphs and partial orderings. *Monatsh. Math.* 73:385–396.
J. W. Alexander
  1939. Ordered sets, complexes and the problem of compactification. *Proc. Nat.*
    Acad. Sci. U.S.A. 25:296–298.
M. Altwegg
  1950. Zur Axiomatik der teilweise geordneten Mengen. *Comment. Math. Helv.*
    24:149–155.
F. W. Anderson and R. L. Blair
  1961. Representations of distributive lattices as lattices of functions. *Math. Ann.*
    143:187–211.
L. W. Anderson
  1961. Locally compact topological lattices. In *Lattice theory*, Proceedings of
    Symposia in Pure Mathematics II, 195–197. Providence, R. I.: American
    Mathematical Society.
G. Ja. Areškin
  1953a. On congruence relations in distributive lattices with zero (in Russian). *Dokl.*
    Akad. Nauk SSSR. N. S. 90:485–486.
  1953b. Free distributive lattices and free bicompact T_0-spaces (in Russian). *Mat.*
    Sbornik. N.S. 33(75):133–156.
S. P. Avann
  1964. Dependence of finiteness conditions in distributive lattices. *Math. Z.* 85:245–
    256.

R. Balbes
  1967. Projective and injective distributive lattices. *Pacific J. Math.* 21:405–420.
  1969. A note on distributive sublattices of a modular lattice. *Fund Math.* 65:219–
    222.

—— and Ph. Dwinger
  a. Coproducts of Boolean algebras and chains with applications to Post
     algebras. *Colloq. Math.* (to appear).
  b. Uniqueness of representation of a distributive lattice as a free product of a
     Boolean algebra and a chain. *Colloq. Math.* (to appear).
—— and G. Grätzer
     Injective and projective Stone algebras. *Duke Math. J.* 38 (1971): 339–347.
—— and A. Horn
1970a. Stone lattices. *Duke Math. J.* 37:537–546.
1970b. Injective and projective Heyting algebras. *Trans. Amer. Math. Soc.* 148:549–
       560.
1970c. Projective distributive lattices. *Pacific J. Math.* 33:273–280.
B. Banaschewski
1970. Injectivity and essential extensions in equational classes of algebras. In
      Proceedings of the conference on universal algebras, pp. 131–147. Kingston,
      Ont.: Queen's University.
E. A. Behrens
1960. Distributiv darstellbare Ringe I. *Math. Z.* 73:409–432.
1961. Distributiv darstellbare Ringe II. *Math. Z.* 76:367–384.
N. D. Belnap and J. H. Spencer
1966. Intensionally complemented distributive lattices. *Portugaliae Math.* 25:99–
      104.
      (See also J. M. Dunn and N. D. Belnap)
G. Birkhoff
1933. On the combination of subalgebras. *Proc. Cambridge Philos. Soc.* 29:441–464.
1935. On the structure of abstract algebras. *Proc. Cambridge Philos. Soc.* 31:433–
      454.
1940. *Lattice theory.* 1st ed. Providence, R.I.: American Mathematical Society.
1944. Subdirect unions in universal algebra. *Bull. Amer. Math. Soc.* 50:764–768.
1946. On groups of automorphisms (in Spanish). *Revista Unión Mat. Argentina.*
      11:155–157.
1948. *Lattice theory.* 2nd ed. Providence, R.I.: American Mathematical Society.
1962. A new interval topology for dually directed sets. *Univ. Nac. Tucumán.*
      Revista Ser. A, 14:325–331.
1967. *Lattice theory.* 3rd ed. Providence, R.I.: American Mathematical Society.
1970. What can lattices do for you? In *Trends in lattice theory*, pp. 1–40. New York:
      Van Nostrand-Reinhold.
R. L. Blair
     (See F. W. Anderson and R. L. Blair)
A. B. Brown
     (See S. Abian and A. B. Brown)
G. Bruns
1959. Verbandstheoretische Kennzeichnung vollständiger Mengenringe. *Arch.*
      Math. (Basel) 10:109–112.
1961. Distributivität und subdirekte Zerlegbarkeit vollständiger Verbände. *Arch.*
      Math. (Basel) 12:61–66.
1962a. Darstellungen und Erweiterungen geordneter Mengen I, II. *J. Reine Angew.*
       Math. 209:167–200; 210:1–23.
1962b. On the representation of Boolean algebras. *Canad Math. Bull.* 5:37–
       41.
1965. Ideal representations of Stone lattices. *Duke Math. J.* 32:555–556.

J. R. Büchi
    1952. Representation of complete lattices by sets. *Portgualiae Math.* 11:151–167.
E. Buttafuoco
    (See R. Musti and E. Buttafuoco)
L. Byrne
    1946. Two brief formulations of Boolean algebra. *Bull. Amer. Math. Soc.* 52:269–272.

A. D. Campbell
    1943. Set-coordinates for lattices. *Bull. Amer. Math. Soc.* 49:395–398.
C. C. Chang and A. Horn
    1961. Prime ideal characterization of generalized Post algebras. In *Lattice theory*, Proceedings of Symposia in Pure Mathematics II, 43–48. Providence, R.I.: American Mathematical Society.
    1962. On the representation of α-complete lattices. *Fund. Math.* 51:253–258.
C. C. Chen and G. Grätzer
    1969a. On the construction of complemented lattices. *J. Algebra.* 11:56–63.
    1969b. Stone lattices I. Construction theorems. *Canad. J. Math.* 21:884–894.
    1969c. Stone lattices II. Structure theorems. *Canad. J. Math.* 21:895–903.
R. Church
    1940. Numerical analysis of certain free distributive structures. *Duke Math. J.* 6:732–734.
I. S. Cohen
    1950. Commutative rings with restricted minimum conditions. *Duke Math. J.* 17:27–42.
M. Cotlar
    1944. A method of construction of structures and its application to topological spaces and abstract arithmetic. *Univ. Nac. Tucumán. Revista Ser. A*, 4:105–157.
H. Crapo and G. C. Rota
    1970. Geometric lattices. In *Trends in lattice theory*, pp. 127–172. New York: Van Nostrand-Reinhold.
P. Crawley
    1962. Regular embeddings which preserve lattice structure. *Proc. Amer. Math. Soc.* 13:748–752.
    (See also R. P. Dilworth and P. Crawley)
H. B. Curry
    1963. *Foundations of mathematical logic.* New York: McGraw-Hill.
M. Curzio
    1953. Alcune limitazioni sul minimo ordine dei reticoli modulari di lunghezza 3 contenenti sottoreticoli d'ordine dato. *Ricerche Mat.* 2:140–147.

A. Daigneault
    1959. Products of polyadic algebras and of their representations. Ph.D. Thesis, Princeton University.
A. C. Davis
    1955. A characterization of complete lattices. *Pacific J. Math.* 5:311–319.
A. Day
    1970. Injectivity in congruence distributive equational classes. Ph.D. Thesis, McMaster University, Hamilton, Ontario.

a. Injectives in non-distributive equational classes of lattices are trivial. *Arch. Math.* (Basel) 21 (1970):113–115.

R. A. Dean and R. H. Oehmke
1964. Idempotent semigroups with distributive right congruence lattices. *Pacific J. Math.* 14:1187–1209.

R. Dedekind
1900. Über die von drei Moduln erzeugte Dualgruppe. *Math. Ann.* 53:371–403.

R. Demarr
1964. Common fixed points for isotone mappings. *Colloq. Math.* 13:45–48.

V. Devidé
1963. On monotonous mappings of complete lattices. *Fund. Math.* 53:147–154.

A. H. Diamond and J. C. C. McKinsey
1947. Algebras and their subalgebras. *Bull. Amer. Math. Soc.* 53:959–962.

R. P. Dilworth
1945. Lattices with unique complements. *Trans. Amer. Math. Soc.* 57:123–154.

1961. Structure and decomposition theory of lattices. In *Lattice theory*, Proceedings of Symposia in Pure Mathematics II, 3–16. Providence, R.I.: American Mathematical Society.

——— and P. Crawley
1960. Decomposition theory for lattices without chain condition. *Trans. Amer. Math. Soc.* 96:1–22.

——— and J. E. McLaughlin
1952. Distributivity in lattices. *Duke Math. J.* 19:683–693.

J. M. Dunn and N. D. Belnap
1968. Homomorphisms of intentionally complemented distributive lattices. *Math. Ann.* 176:28–38.

Ph. Dwinger
(See R. Balbes and Ph. Dwinger).

R. Engelking
1965. Cartesian products and dyadic spaces. *Fund. Math.* 57:287–304.

G. Epstein
1960. The lattice theory of Post algebras. *Trans. Amer. Math. Soc.* 95:300–317.

N. D. Filippov
1966. Projectivity of lattices (in Russian). *Mat. Sb.* 70(112):36–54.

P. A. Freĭdman
1967. Rings with a distributive lattice of subrings (in Russian). *Mat. Sbornik* 73(4): 513–534.

H. Friedman and D. Tamari
1967. Problèmes d'associativité: une structure de treillis finis induite par une loi demi-associative. *J. Combinatorial Theory* 2:215–242.

O. Frink
1942. Topology in lattices. *Trans. Amer. Math. Soc.* 51:569–582.

1962. Pseudo-complements in semi-lattices. *Duke Math. J.* 29:505–514.

1964. Compactifications and semi-normal spaces. *Amer. J. Math.* 86:602–607.

R. Frucht
1948. On the construction of partially ordered systems with a given group of automorphisms. *Revista Unión Mat. Argentina* 13:12–18.

1950. Lattices with a given abstract group of automorphisms. *Canad. J. Math.* 2:417–419.

L. Fuchs
1949. Über die ideale aritmetischer Ringe. *Comm. Math. Helv.* 23:334–341.

N. Funayama
1944. On the completion by cuts of distributive lattices. *Proc. Imp. Acad. Tokyo* 20:1–2.
1953. Notes on lattice theory IV. On partial (semi-) lattices. *Bull. Yamagata Univ.* (Nat. Sci.) 2:171–184.
1959. Imbedding infinitely distributive lattices completely isomorphically into Boolean algebras. *Nagoya Math. J.* 15:71–81.
————— and T. Nakayama
1942. On the distributivity of a lattice of lattice-congruences. *Proc. Imp. Acad. Tokyo* 18:553–554.

F. Galvin and B. Jónsson
1961. Distributive sublattices of a free lattice. *Canad. J. Math.* 13:265–272.

A. Ghouilà-Houri
1962. Caractérisation des graphes non orientés dont on peut orienter les arêtes de manière à obtenir le graphe d'une relation d'ordre. *C. R. Acad. Sci. Paris.* Sér. A–B, 254:1370–1371.

E. N. Gilbert
1954. Lattice theoretic properties of frontal switching functions. *J. Math. Phys.* 33:57–67.

L. Gillman and M. Jerison
1960. *Rings of continuous functions.* University Series in Higher Mathematics. Princeton, N.J.: Van Nostrand.

P. C. Gilmore and A. J. Hoffman
1964. A characterization of comparability graphs and of interval graphs. *Canad. J. Math.* 16:539–548.

V. Glivenko
1929. Sur quelques points de la logique de M. Brouwer. *Bull Acad. des Sci. de Belgique* 15:183–188.

M. M. Gluhov
1960. On the problem of isomorphism of lattices (in Russian). *Dokl. Akad. Nauk SSSR.* 132:254–256 (English translation: *Soviet Math. Dokl.* I:519–522).

G. Grätzer
1959. Standard ideals (in Hungarian). *Magyar Tud. Akad. Mat. Fiz. Oszt. Közl.* 9:81–97.
1962. On Boolean functions (Notes on lattice theory II). *Rev. Roumaine Math. Pures Appl.* 7:693–697.
1963. A generalization of Stone's representation theorem for Boolean algebras. *Duke J. Math.* 30:469–474.
1964. Boolean functions on distributive lattices. *Acta Math. Acad. Sci. Hungar.* 15:195–201.
1966. Equational classes of lattices. *Duke J. Math.* 33:613–622.
1968. *Universal algebra.* University Series in Higher Mathematics. Princeton, N.J.: Van Nostrand.
1969. Stone algebras form an equational class (Notes on lattice theory III). *J. Austral. Math. Soc.* 9:308–309.
1970. Universal algebra. In *Trends in lattice theory*, pp. 173–215. New York: Van Nostrand-Reinhold.

a. A reduced free product of lattices. *Fund. Math.* (to appear).

(See also R. Balbes and G. Grätzer; C. C. Chen and G. Grätzer)

———— and H. Lakser

1968. Extension theorems on congruences of partial lattices. Abstract. *Notices Amer. Math. Soc.* 15:732, 785.

1969a. Chain conditions in distributive free product of lattices. *Trans. Amer. Math. Soc.* 144:301–312.

1969b. Some applications of free distributive products. Abstract. *Notices Amer. Math. Soc.* 16:405.

a. The structure of pseudo-complemented distributive lattices II. Congruence extension and amalgamation. *Trans. Amer. Math. Soc.* 156 (1971):343–358.

b. The structure of pseudo-complemented distributive lattices III. Injectives and absolute subretracts (to appear).

c. Some new relations on operators in general, and for pseudo-complemented distributive lattices in particular. Abstract. *Notices Amer. Math. Soc.* 17:642.

————, H. Lakser, and C. R. Platt

1970. Free products of lattices. *Fund. Math.* 69:233–240.

———— and R. N. McKenzie

1967. Equational spectra and reduction of identities. Abstract. *Notices Amer. Math. Soc.* 14:697.

———— and E. T. Schmidt

1957a. On the Jordan-Dedekind chain condition. *Acta Sci. Math.* (Szeged) 18:52–56.

1957b. On a problem of M. H. Stone. *Acta Math. Acad. Sci. Hungar.* 8:455–460.

1958a. On the lattice of all join-endomorphisms of a lattice. *Proc. Amer. Math. Soc.* 9:722–726.

1958b. Characterizations of relatively complemented distributive lattices. *Publ. Math. Debrecen* 5:275–287.

1958c. Two notes on lattice-congruences. *Ann. Univ. Sci. Budapest. Eötvös Sect. Math.* 1:83–87.

1958d. On ideal theory for lattices. *Acta Sci. Math.* (Szeged) 19:82–92.

1958e. Ideals and congruence relations in lattices. *Acta Math. Acad. Sci. Hungar.* 9:137–175.

1958f. On the generalized Boolean algebra generated by a distributive lattice. *Indag. Math.* 20:547–553.

1961. Standard ideals in lattices. *Acta Math. Acad. Sci. Hungar.* 12:17–86.

1962. On congruence lattices of lattices. *Acta Math. Acad. Sci. Hungar.* 13:179–185.

———— and J. Sichler

a. Endomorphism semigroups (and categories) of bounded lattices. *Pacific J. Math.* 36 (1971):639–647.

———— and B. Wolk

1970. Finite projective distributive lattices. *Canad. Math. Bull.* 13:139–140.

P. R. Halmos

1961. Injective and projective Boolean algebras. In *Lattice theory*, Proceedings of Symposia in Pure Mathematics II, 114–122. Providence, R.I.: American Mathematical Society.

1963. *Lectures on Boolean algebras*. Mathematical Studies No.1. Princeton, N.J.: Van Nostrand.

G. Hansel
1967. Problèmes de dénombrement et d'évaluation de bornes concernant les
éléments du treillis distributif libre. *Publ. Inst. Statist. Univ. Paris* 16:159–
218; and 16:219–300.
M. A. Harrison
1965. *Introduction to switching and automata theory.* New York: McGraw-Hill.
J. Hashimoto
1952. Ideal theory for lattices. *Math. Japon.* 2:149–186.
(See also S. Kinugawa and J. Hashimoto)
G. Havas and M. Ward
1969. Lattices with sublattices of a given order. *J. Combinatorial Theory* 7:281–282.
D. Higgs
1971. Lattices isomorphic to their ideal lattices. *Algebra Universalis* 1:71–72
A. J. Hoffman
(See R. C. Gilmore and A. J. Hoffman)
A. Horn
1962. On α-homomorphic images of α-rings of sets. *Fund. Math.* 51:259–266.
1968. A property of free Boolean algebras. *Proc. Amer. Math. Soc.* 19:142–143.
1969a. Logic with truth values in a linearly ordered Heyting algebra. *J. Symbolic
Logic* 43:395–408.
1969b. Free L-algebras. *J. Symbolic Logic* 34:475–480.
(See also R. Balbes and A. Horn; C. C. Chang and A. Horn)
S. Huang and D. Tamari
a. Problems of associativity: A simple proof for the lattice property of systems
ordered by a semi-associative law. *J. Combinatorial Theory* (to appear).
E. V. Huntington
1904. Sets of independent postulates for the algebra of logic. *Trans. Amer. Math.
Soc.* 5:288–309.
T. Iwamura
1944. A lemma on directed sets (in Japanese). *Zenkoku Shijo Sugaku Danwakai*
262: 107–111.
J. Jakubík
1954a. On lattices whose graphs are isomorphic (in Russian). *Czechoslovak Math. J.*
4(79):131–141.
1954b. On the graph isomorphism of semi-modular lattices (in Slovak). *Mat. Časopis
Sloven. Akad. Vied.* 4:162–177.
1957. Remark on the Jordan-Dedekind condition in Boolean algebras (in Slovak).
Časopis Pěst. Mat. 82:44–46.
1958. On chains in Boolean lattices (in Slovak). *Mat. Časopis Sloven. Akad. Vied.*
8:193–202.
——— and M. Kolibiar
1954. On some properties of a pair of lattices (in Russian). *Czechoslovak Math. J.*
4(79):1–27.
Ch. U. Jensen
1963. On characterisations of Prüfer rings. *Math. Scand.* 13:190–198.
M. Jerison
(See L. Gillman and M. Jerison)
B. Jónsson
1951. A Boolean algebra without proper automorphisms. *Proc. Amer. Math. Soc.*
2:766–770.

1955. Distributive sublattices of a modular lattice. *Proc. Amer. Math. Soc.* 6:682–688.
1956. Universal relational systems. *Math. Scand.* 4:193–208.
1960. Homogeneous universal relational systems. *Math. Scand.* 8:137–142.
1961. Sublattices of a free lattice. *Canad. J. Math.* 13:256–264.
1965. Extensions of relational structures. *Proc. of the 1963 Int. Symp. at Berkeley*, 146–157. Amsterdam: North Holland Publishing Co.
1967. Algebras whose congruence lattices are distributive. *Math. Scand.* 21:110–121.
(See also F. Galvin and B. Jónsson)

J. A. Kalman
1968. Two axiom definition for lattices. *Rev. Roumaine Math. Pures Appl.* 13:669–670.

I. Kaplansky
1947. Lattices of continuous functions. *Bull. Amer. Math. Soc.* 53:617–623.

D. A. Kappos and F. Papangelou
1966. Remarks on the extension of continuous lattices. *Math. Ann.* 166:277–283.

M. Katětov
1951. Remarks on Boolean algebras. *Colloq. Math.* 2:229–235.

T. Katriňák
1966. Notes on Stone lattices I (in Russian). *Mat. Časopis Sloven. Akad. Vied.* 16:128–142.
1967. Notes on Stone lattices II (in Russian). *Mat. Časopis Sloven. Akad. Vied.* 17:20–37.
1970. Die Kennzeichnung der distributiven pseudokomplementären Halbverbände. *J. Reine Angew. Math.* 241:160–179.

S. Kinugawa and J. Hashimoto
1966. On relative maximal ideals in lattices. *Proc. Japan Acad.* 42:1–4.

D. Kleitman
1969. On Dedekind's problem: The number of monotone Boolean functions. *Proc. Amer. Math. Soc.* 21:677–682.

S. R. Kogalovskiĭ
1964. On linearly complete ordered sets (in Russian). *Uspehi Mat. Nauk.* 19, No. 2, 147–150.

M. Kolibiar
1956. On the axiomatic of modular lattices (in Russian). *Czechoslovak Math. J.* 6(81):381–386.
(See also J. Jakubik and M. Kolibiar)

A. Komatu
1943. On a characterization of join homomorphic transformation-lattice. *Proc. Imp. Acad. Tokyo* 19:119–124.

H. Lakser
1970. Injective hulls of Stone algebras. *Proc. Amer. Math. Soc.* 24:524–529.
   a. The structure of pseudo-complemented distributive lattices I. Subdirect decomposition. *Trans. Amer. Math. Soc.* 156 (1971):335–342.
(See also G. Grätzer and H. Lakser; G. Grätzer, H. Lakser, and C. R. Platt)

K. B. Lee
1970. Equational classes of distributive pseudo-complemented lattices. *Canad. J. Math.* 22:881–891.

H. F. J. Lowig
1943. On the importance of the relation [(A, B), (A, C)] < [(B, C), (C, A), (A, B)] between three elements of a structure. *Ann. of Math.* (2)44:573-579.
H. M. MacNeille
1937. Partially ordered sets. *Trans. Amer. Math. Soc.* 42:416-460.
1939. Extension of a distributive lattice to a Boolean ring. *Bull. Amer. Math. Soc.* 45:452-455.
M. Mandelker
1970. Relative annihilators in lattices. *Duke Math. J.* 37:377-386.
J. M. Maranda
1964. Injective structures. *Trans. Amer. Math. Soc.* 110:98-135.
R. N. McKenzie
  a. Equational bases for lattice theories. *Math. Scand.* 27: 24-38.
  b. Equational bases and non-modular lattice varieties. *Trans. Amer. Math. Soc.* (to appear).
  (See also G. Grätzer and R. N. McKenzie)
—— and J. Sichler
  a. Endomorphisms of finite complemented lattices and of lattices of finite length (to appear).
J. C. C. McKinsey
  (See A. H. Diamond and J. C. C. McKinsey)
J. E. McLaughlin
1961. The normal completion of a complemented modular point lattice. In *Lattice theory*, Proceedings of Symposia in Pure Mathematics II, 78-80. Providence, R.I.: American Mathematical Society.
  (See also R. P. Dilworth and J. E. McLaughlin)
G. Michler and R. Wille
1970. Die primitiven Klassen arithmetischer Ringe. *Math. Z.* 113:369-372.
B. Mitchell
1965. *Theory of categories.* New York: Academic Press.
A. Monteiro
1947. Sur l'arithmétique des filtres premiers. *C. R. Acad. Sci. Paris* 225:846-848.
1955. Axiomes indépendants pour les algèbres de Brouwer. *Revista Unión Mat. Argentina* 17:149-160.
M. D. Morley and R. L. Vaught
1962. Homogeneous universal models. *Math. Scand.* 11:37-57.
M. Mostowski and A. Tarski
1939. Boolesche Ringe mit geordneter Basis. *Fund. Math.* 32:69-86.
R. Musti and E. Buttafuoco
1956. Sui subreticoli distributivi dei reticoli modulari. *Boll. Un. Mat. Ital.* (3)11: 584-587.

L. Nachbin
1947. Une propriété carastéristique des algèbres booléiennes. *Portugaliae Math.* 6:115-118.
1949. On a characterization of the lattice of all ideals of a Boolean ring. *Fund. Math.* 36:137-142.
T. Nakayama
  (See N. Funayama and T. Nakayama)
W. C. Nemitz
1965. Implicative semi-lattices. *Trans. Amer. Math. Soc.* 117:128-142.

202 BIBLIOGRAPHY

A. Nerode
1959. Some Stone spaces and recursion theory. *Duke Math. J.* 26:397–406.
J. von Neumann
1936. *Lectures on continuous geometries, Princeton, 1936–1937* (mimeographed). Reprinted as *Continuous geometry.* Princeton, N.J.: Princeton University Press, 1960.
E. Noether
1927. Abstrakter Aufbau der Idealtheorie in algebraischen Zahl- und Funktionenkörpern. *Math. Annalen* 96:26–61.

R. H. Oehmke
(See R. A. Dean and R. H. Oehmke)
O. Ore
1937. Structures and group theory I. *Duke Math. J.* 3:149–174.
1938. Structures and group theory II. *Duke Math. J.* 4:247–269.
1940. Remarks on the structures and group relations. *Vierteljschr. Naturforsch. Ges. Zürich,* 85 Beiblatt (Festschrift Rudolf Fueter), 1–4.

R. Padmanabhan
1966. On axioms for semilattices. *Canad. Math. Bull.* 9:357–358.
1968. A note on Kalman's paper. *Rev. Roumaine Math. Pures Appl.* 13:1149–1152.
1969. Two identities for lattices. *Proc. Amer. Math. Soc.* 20:409–412.
F. Papangelou
(See D. A. Kappos and F. Papangelou)
D. Papert
1964. Congruence relations in semilattices. *J. London Math. Soc.* 39:723–729.
A. Pelczar
1961. On the invariant points of a transformation. *Ann. Polon. Math.* 11:199–202.
1962. On the extremal solutions of a functional equation. *Zeszyty Nauk. Uniw. Jagiello. Prace Mat.* No. 7:9–11.
R. Permutti
1964. Sulle semicongruenze di un reticolo. *Rend. Accad. Sci. Fis. Mat. Napoli.* (4)31:160–167.
R. S. Pierce
1961. Some questions about complete Boolean algebras. In *Lattice theory,* Proceedings of Symposia in Pure Mathematics II, 129–140. Providence, R.I.: American Mathematical Society.
1968. *Introduction to the theory of abstract algebras.* New York: Holt, Rinehart and Winston.
a. Topological Boolean Algebras. In *Proceedings of the conference on universal algebra,* pp. 107–130. Kingston, Ont.: Queen's University, 1970.
C. R. Platt
(See G. Grätzer, H. Lakser, and C. R. Platt)
E. L. Post
1921. Introduction to a general theory of elementary propositions. *Amer. J. Math.* 43:163–185.
D. H. Potts
1965. Axioms for semilattices. *Canad. Math. Bull.* 8:519.

G. N. Raney
1952. Completely distributive complete lattices. *Proc. Amer. Math. Soc.* 3:677–680.

1953. A subdirect-union representation for completely distributive complete lattices. *Proc. Amer. Math. Soc.* 4:518–522.

H. Rasiowa and R. Sikorski

1963. The mathematics of metamathematics. *Monog. Mat.* Vol. XLI. Warsaw: Państowe Wydawn. Naukowe.

B. C. Rennie

1951. *The theory of lattices.* Cambridge, England: Forster and Jagg.

I. Reznikoff

1963. Chaînes de formules. *C. R. Acad. Sci. Paris* 256:5021–5023.

P. Ribenboim

1949. Characterization of the sup-complement in a distributive lattice with last element. *Summa Brasil. Math.* 2, No. 4, 43–49.

J. Riečan

1958. To the axiomatics of modular lattices (in Slovak). *Acta Fac. Rerum Natur. Univ. Comenian. Math.* 2:257–262.

L. Rieger

1949. A note on topological representations of distributive lattices. *Časopis Pěst. Mat.* 74:55–61.

1951. Some remarks on automorphisms of Boolean algebras. *Fund. Math.* 38:209–216.

H. L. Rolf

1958. The free lattice generated by a set of chains. *Pacific J. Math.* 8:585–595.

P. C. Rosenbloom

1942. Post algebras I. Postulates and general theory. *Amer. J. Math.* 64:167–188.

G. C. Rota

(See H. Crapo and G. C. Rota)

S. Rudeanu

1963. *Axioms for lattices and Boolean algebras* (in Roumanian). Bucharest: Editura Academiei Republicii Populare Romîne.

D. Sachs

1962. The lattice of subalgebras of a Boolean algebra. *Canad. J. Math.* 14:451–460.

Y. Sampei

1953. On lattice completions and closure operations. *Comment. Math. Univ. St. Paul* 2:55–70; and 3 (1954): 29–30.

E. T. Schmidt

1962. Über die Kongruenzverbände der Verbände. *Publ. Math. Debrecen* 9:243–256.

1968. Zur Charakterisierung der Kongruenzverbände der Verbände. *Mat. Časopis Sloven. Akad. Vied.* 18:3–20.

1969. *Kongruenzrelationen algebraischer Strukturen.* Berlin: VEB Deutscher Verlag der Wissenschaften.

(See also G. Grätzer and E. T. Schmidt)

L. N. Ševrin

1964. Projectivities of semi-lattices (in Russian). *Dokl. Akad. Nauk SSSR.* 154:538–541.

M. Sholander

1951. Postulates for distributive lattices. *Canad. J. Math.* 3:28–30.

J. Sichler
  a. Every monoid is isomorphic to the monoid of all non-constant endo-
     morphisms of a lattice (to appear).
     (See also G. Grätzer and J. Sichler; R. N. McKenzie and J. Sichler)
R. Sikorski
  1964. *Boolean algebras.* 2nd ed. Ergebnisse der Mathematik und ihrer Grenz-
     gebiete, Neue Folge, Band 25. New York: Academic Press. Berlin: Springer
     Verlag.
     (See also H. Rasiowa and R. Sikorski)
F. M. Sioson
  1964. Equational bases of Boolean algebras. *J. Symbolic Logic.* 29:115–124.
Ju. I. Sorkin
  1951. Independent systems of axioms defining a lattice (in Russian). *Ukrain. Mat.
     Ž.* 3:85–97.
  1952. Free unions of lattices (in Russian). *Mat. Sb.* 30(72):677–694.
T. P. Speed
  1969. On Stone lattices. *J. Austral. Math. Soc.* 9:293–307.
J. H. Spencer
     (See N. D. Belnap and J. H. Spencer)
A. K. Steiner
  1966. The lattice of topologies: Structure and complementation. *Trans. Amer. Math.
     Soc.* 122:379–398.
M. H. Stone
  1936. The theory of representations for Boolean algebras. *Trans. Amer. Math.
     Soc.* 40:37–111.
  1937. Topological representations of distributive lattices and Brouwerian logics.
     Časopis Pěst. Mat. 67:1–25.

K. Takeuchi
  1951. On maximal proper sublattices. *J. Math. Soc. Japan.* 2:228–230.
D. Tamari
  1951. Monoides préordonnés et chaînes de Malcev. Thèse, Université de Paris.
     (See also H. Friedman and D. Tamari; S. Huang and D. Tamari)
A. Tarski
  1930. Sur les classes d'ensembles closes par rapport à certaines opérations élémen-
     taires. *Fund. Math.* 16:181–304.
  1955. A lattice-theoretical fixpoint theorem and its applications. *Pacific J. Math.*
     5:285–309.
  1968. Equational logic and equational theories of algebras. In *Contributions to
     mathematical logic*, pp. 275–288. Amsterdam: North Holland Publishing Co.
     (See also M. Mostowski and A. Tarski)
T. Traczyk
  1963. Axioms and some properties of Post algebras. *Colloq. Math.* 10:193–209.
  1967. On Post algebras with uncountable chain of constants. *Bull. Acad. Polon. Sci.
     Sér. Sci. Math. Astronom. Phys.* 15:673–680.

R. Vaidyanathaswamy
  1947. *Set topology.* 1st ed. Madras.
  1960. *Set topology.* 2d ed. New York: Chelsea.
J. Varlet
  1963. Contribution à l'étude des treillis pseudo-complémentés et des treillis de
     Stone. *Mém. Soc. Roy. Sci. Liège Coll. in –8°,* Vol. 8, No. 4.

1965. Congruence dans les demi-lattis. *Bull. Soc. Roy. Sci. Liège.* 34:231–240.
1966. On the characterization of Stone lattices. *Acta Sci. Math.* (Szeged). 27:81–84.
1968. Algèbres de Lukasiewicz trivalentes. *Bull. Soc. Royale Sci. Liège.* 36:399–408.

R. L. Vaught
(See M. D. Morley and R. L. Vaught)

H. Wallman
1938. Lattices and topological spaces. *Ann. of Math.* 39:112–126.

M. Ward
(See G. Havas and M. Ward)

H. Werner and R. Wille
1970. Characterisierungen der primitiven Klassen arithmetischer Ringe. *Math. Z.* 115:197–200.

P. M. Whitman
1943. Splittings of a lattice. *Amer. J. Math.* 65:179–196.

R. Wille
(See G. Michler and R. Wille; H. Werner and R. Wille)

B. Wolk
(See G. Grätzer and B. Wolk)

E. S. Wolk
1957. Dedekind completeness and a fixed-point theorem. *Canad. J. Math.* 9:400–405.

G. Zacher
1952. Sugli emimorfismi superiori ed inferiori tra reticoli. *Rend Accad. Sci. Fis. Mat. Napoli* (4)19:45–56.

INDEX

A CATALOG OF SELECTED
DOVER BOOKS
IN SCIENCE AND MATHEMATICS

Astronomy

BURNHAM'S CELESTIAL HANDBOOK, Robert Burnham, Jr. Thorough guide to the stars beyond our solar system. Exhaustive treatment. Alphabetical by constellation: Andromeda to Cetus in Vol. 1; Chamaeleon to Orion in Vol. 2; and Pavo to Vulpecula in Vol. 3. Hundreds of illustrations. Index in Vol. 3. 2,000pp. 6⅛ x 9¼.
Vol. I: 0-486-23567-X
Vol. II: 0-486-23568-8
Vol. III: 0-486-23673-0

EXPLORING THE MOON THROUGH BINOCULARS AND SMALL TELESCOPES, Ernest H. Cherrington, Jr. Informative, profusely illustrated guide to locating and identifying craters, rills, seas, mountains, other lunar features. Newly revised and updated with special section of new photos. Over 100 photos and diagrams. 240pp. 8¼ x 11. 0-486-24491-1

THE EXTRATERRESTRIAL LIFE DEBATE, 1750–1900, Michael J. Crowe. First detailed, scholarly study in English of the many ideas that developed from 1750 to 1900 regarding the existence of intelligent extraterrestrial life. Examines ideas of Kant, Herschel, Voltaire, Percival Lowell, many other scientists and thinkers. 16 illustrations. 704pp. 5⅜ x 8½. 0-486-40675-X

THEORIES OF THE WORLD FROM ANTIQUITY TO THE COPERNICAN REVOLUTION, Michael J. Crowe. Newly revised edition of an accessible, enlightening book recreates the change from an earth-centered to a sun-centered conception of the solar system. 242pp. 5⅜ x 8½. 0-486-41444-2

A HISTORY OF ASTRONOMY, A. Pannekoek. Well-balanced, carefully reasoned study covers such topics as Ptolemaic theory, work of Copernicus, Kepler, Newton, Eddington's work on stars, much more. Illustrated. References. 521pp. 5⅜ x 8½.
0-486-65994-1

A COMPLETE MANUAL OF AMATEUR ASTRONOMY: TOOLS AND TECHNIQUES FOR ASTRONOMICAL OBSERVATIONS, P. Clay Sherrod with Thomas L. Koed. Concise, highly readable book discusses: selecting, setting up and maintaining a telescope; amateur studies of the sun; lunar topography and occultations; observations of Mars, Jupiter, Saturn, the minor planets and the stars; an introduction to photoelectric photometry; more. 1981 ed. 124 figures. 25 halftones. 37 tables. 335pp. 6½ x 9¼. 0-486-40675-X

AMATEUR ASTRONOMER'S HANDBOOK, J. B. Sidgwick. Timeless, comprehensive coverage of telescopes, mirrors, lenses, mountings, telescope drives, micrometers, spectroscopes, more. 189 illustrations. 576pp. 5⅜ x 8¼. (Available in U.S. only.)
0-486-24034-7

STARS AND RELATIVITY, Ya. B. Zel'dovich and I. D. Novikov. Vol. 1 of *Relativistic Astrophysics* by famed Russian scientists. General relativity, properties of matter under astrophysical conditions, stars, and stellar systems. Deep physical insights, clear presentation. 1971 edition. References. 544pp. 5⅜ x 8¼. 0-486-69424-0

Chemistry

THE SCEPTICAL CHYMIST: THE CLASSIC 1661 TEXT, Robert Boyle. Boyle defines the term "element," asserting that all natural phenomena can be explained by the motion and organization of primary particles. 1911 ed. viii+232pp. 5⅜ x 8½.
0-486-42825-7

RADIOACTIVE SUBSTANCES, Marie Curie. Here is the celebrated scientist's doctoral thesis, the prelude to her receipt of the 1903 Nobel Prize. Curie discusses establishing atomic character of radioactivity found in compounds of uranium and thorium; extraction from pitchblende of polonium and radium; isolation of pure radium chloride; determination of atomic weight of radium; plus electric, photographic, luminous, heat, color effects of radioactivity. ii+94pp. 5⅜ x 8½.   0-486-42550-9

CHEMICAL MAGIC, Leonard A. Ford. Second Edition, Revised by E. Winston Grundmeier. Over 100 unusual stunts demonstrating cold fire, dust explosions, much more. Text explains scientific principles and stresses safety precautions. 128pp. 5⅜ x 8½.   0-486-67628-5

THE DEVELOPMENT OF MODERN CHEMISTRY, Aaron J. Ihde. Authoritative history of chemistry from ancient Greek theory to 20th-century innovation. Covers major chemists and their discoveries. 209 illustrations. 14 tables. Bibliographies. Indices. Appendices. 851pp. 5⅜ x 8½.   0-486-64235-6

CATALYSIS IN CHEMISTRY AND ENZYMOLOGY, William P. Jencks. Exceptionally clear coverage of mechanisms for catalysis, forces in aqueous solution, carbonyl- and acyl-group reactions, practical kinetics, more. 864pp. 5⅜ x 8½.
0-486-65460-5

ELEMENTS OF CHEMISTRY, Antoine Lavoisier. Monumental classic by founder of modern chemistry in remarkable reprint of rare 1790 Kerr translation. A must for every student of chemistry or the history of science. 539pp. 5⅜ x 8½.   0-486-64624-6

THE HISTORICAL BACKGROUND OF CHEMISTRY, Henry M. Leicester. Evolution of ideas, not individual biography. Concentrates on formulation of a coherent set of chemical laws. 260pp. 5⅜ x 8½.   0-486-61053-5

A SHORT HISTORY OF CHEMISTRY, J. R. Partington. Classic exposition explores origins of chemistry, alchemy, early medical chemistry, nature of atmosphere, theory of valency, laws and structure of atomic theory, much more. 428pp. 5⅜ x 8½. (Available in U.S. only.)   0-486-65977-1

GENERAL CHEMISTRY, Linus Pauling. Revised 3rd edition of classic first-year text by Nobel laureate. Atomic and molecular structure, quantum mechanics, statistical mechanics, thermodynamics correlated with descriptive chemistry. Problems. 992pp. 5⅜ x 8½.   0-486-65622-5

FROM ALCHEMY TO CHEMISTRY, John Read. Broad, humanistic treatment focuses on great figures of chemistry and ideas that revolutionized the science. 50 illustrations. 240pp. 5⅜ x 8½.   0-486-28690-8

Engineering

DE RE METALLICA, Georgius Agricola. The famous Hoover translation of greatest treatise on technological chemistry, engineering, geology, mining of early modern times (1556). All 289 original woodcuts. 638pp. 6¾ x 11.   0-486-60006-8

FUNDAMENTALS OF ASTRODYNAMICS, Roger Bate et al. Modern approach developed by U.S. Air Force Academy. Designed as a first course. Problems, exercises. Numerous illustrations. 455pp. 5⅜ x 8½.   0-486-60061-0

DYNAMICS OF FLUIDS IN POROUS MEDIA, Jacob Bear. For advanced students of ground water hydrology, soil mechanics and physics, drainage and irrigation engineering and more. 335 illustrations. Exercises, with answers. 784pp. 6⅛ x 9¼.
0-486-65675-6

THEORY OF VISCOELASTICITY (Second Edition), Richard M. Christensen. Complete consistent description of the linear theory of the viscoelastic behavior of materials. Problem-solving techniques discussed. 1982 edition. 29 figures. xiv+364pp. 6⅛ x 9¼.   0-486-42880-X

MECHANICS, J. P. Den Hartog. A classic introductory text or refresher. Hundreds of applications and design problems illuminate fundamentals of trusses, loaded beams and cables, etc. 334 answered problems. 462pp. 5⅜ x 8½.   0-486-60754-2

MECHANICAL VIBRATIONS, J. P. Den Hartog. Classic textbook offers lucid explanations and illustrative models, applying theories of vibrations to a variety of practical industrial engineering problems. Numerous figures. 233 problems, solutions. Appendix. Index. Preface. 436pp. 5⅜ x 8½.   0-486-64785-4

STRENGTH OF MATERIALS, J. P. Den Hartog. Full, clear treatment of basic material (tension, torsion, bending, etc.) plus advanced material on engineering methods, applications. 350 answered problems. 323pp. 5⅜ x 8½.   0-486-60755-0

A HISTORY OF MECHANICS, René Dugas. Monumental study of mechanical principles from antiquity to quantum mechanics. Contributions of ancient Greeks, Galileo, Leonardo, Kepler, Lagrange, many others. 671pp. 5⅜ x 8½. 0-486-65632-2

STABILITY THEORY AND ITS APPLICATIONS TO STRUCTURAL MECHANICS, Clive L. Dym. Self-contained text focuses on Koiter postbuckling analyses, with mathematical notions of stability of motion. Basing minimum energy principles for static stability upon dynamic concepts of stability of motion, it develops asymptotic buckling and postbuckling analyses from potential energy considerations, with applications to columns, plates, and arches. 1974 ed. 208pp. 5⅜ x 8½.
0-486-42541-X

METAL FATIGUE, N. E. Frost, K. J. Marsh, and L. P. Pook. Definitive, clearly written, and well-illustrated volume addresses all aspects of the subject, from the historical development of understanding metal fatigue to vital concepts of the cyclic stress that causes a crack to grow. Includes 7 appendixes. 544pp. 5⅜ x 8½. 0-486-40927-9

ROCKETS, Robert Goddard. Two of the most significant publications in the history of rocketry and jet propulsion: "A Method of Reaching Extreme Altitudes" (1919) and "Liquid Propellant Rocket Development" (1936). 128pp. 5⅜ x 8½. 0-486-42537-1

STATISTICAL MECHANICS: PRINCIPLES AND APPLICATIONS, Terrell L. Hill. Standard text covers fundamentals of statistical mechanics, applications to fluctuation theory, imperfect gases, distribution functions, more. 448pp. 5⅜ x 8½. 0-486-65390-0

ENGINEERING AND TECHNOLOGY 1650–1750: ILLUSTRATIONS AND TEXTS FROM ORIGINAL SOURCES, Martin Jensen. Highly readable text with more than 200 contemporary drawings and detailed engravings of engineering projects dealing with surveying, leveling, materials, hand tools, lifting equipment, transport and erection, piling, bailing, water supply, hydraulic engineering, and more. Among the specific projects outlined-transporting a 50-ton stone to the Louvre, erecting an obelisk, building timber locks, and dredging canals. 207pp. 8⅜ x 11¼. 0-486-42232-1

THE VARIATIONAL PRINCIPLES OF MECHANICS, Cornelius Lanczos. Graduate level coverage of calculus of variations, equations of motion, relativistic mechanics, more. First inexpensive paperbound edition of classic treatise. Index. Bibliography. 418pp. 5⅜ x 8½. 0-486-65067-7

PROTECTION OF ELECTRONIC CIRCUITS FROM OVERVOLTAGES, Ronald B. Standler. Five-part treatment presents practical rules and strategies for circuits designed to protect electronic systems from damage by transient overvoltages. 1989 ed. xxiv+434pp. 6⅛ x 9¼. 0-486-42552-5

ROTARY WING AERODYNAMICS, W. Z. Stepniewski. Clear, concise text covers aerodynamic phenomena of the rotor and offers guidelines for helicopter performance evaluation. Originally prepared for NASA. 537 figures. 640pp. 6⅛ x 9¼. 0-486-64647-5

INTRODUCTION TO SPACE DYNAMICS, William Tyrrell Thomson. Comprehensive, classic introduction to space-flight engineering for advanced undergraduate and graduate students. Includes vector algebra, kinematics, transformation of coordinates. Bibliography. Index. 352pp. 5⅜ x 8½. 0-486-65113-4

HISTORY OF STRENGTH OF MATERIALS, Stephen P. Timoshenko. Excellent historical survey of the strength of materials with many references to the theories of elasticity and structure. 245 figures. 452pp. 5⅜ x 8½. 0-486-61187-6

ANALYTICAL FRACTURE MECHANICS, David J. Unger. Self-contained text supplements standard fracture mechanics texts by focusing on analytical methods for determining crack-tip stress and strain fields. 336pp. 6⅛ x 9¼. 0-486-41737-9

STATISTICAL MECHANICS OF ELASTICITY, J. H. Weiner. Advanced, self-contained treatment illustrates general principles and elastic behavior of solids. Part 1, based on classical mechanics, studies thermoelastic behavior of crystalline and polymeric solids. Part 2, based on quantum mechanics, focuses on interatomic force laws, behavior of solids, and thermally activated processes. For students of physics and chemistry and for polymer physicists. 1983 ed. 96 figures. 496pp. 5⅜ x 8½. 0-486-42260-7

Mathematics

FUNCTIONAL ANALYSIS (Second Corrected Edition), George Bachman and Lawrence Narici. Excellent treatment of subject geared toward students with background in linear algebra, advanced calculus, physics and engineering. Text covers introduction to inner-product spaces, normed, metric spaces, and topological spaces; complete orthonormal sets, the Hahn-Banach Theorem and its consequences, and many other related subjects. 1966 ed. 544pp. 6⅛ x 9¼. 0-486-40251-7

ASYMPTOTIC EXPANSIONS OF INTEGRALS, Norman Bleistein & Richard A. Handelsman. Best introduction to important field with applications in a variety of scientific disciplines. New preface. Problems. Diagrams. Tables. Bibliography. Index. 448pp. 5⅜ x 8½. 0-486-65082-0

VECTOR AND TENSOR ANALYSIS WITH APPLICATIONS, A. I. Borisenko and I. E. Tarapov. Concise introduction. Worked-out problems, solutions, exercises. 257pp. 5⅜ x 8¼. 0-486-63833-2

AN INTRODUCTION TO ORDINARY DIFFERENTIAL EQUATIONS, Earl A. Coddington. A thorough and systematic first course in elementary differential equations for undergraduates in mathematics and science, with many exercises and problems (with answers). Index. 304pp. 5⅜ x 8½. 0-486-65942-9

FOURIER SERIES AND ORTHOGONAL FUNCTIONS, Harry F. Davis. An incisive text combining theory and practical example to introduce Fourier series, orthogonal functions and applications of the Fourier method to boundary-value problems. 570 exercises. Answers and notes. 416pp. 5⅜ x 8½. 0-486-65973-9

COMPUTABILITY AND UNSOLVABILITY, Martin Davis. Classic graduate-level introduction to theory of computability, usually referred to as theory of recurrent functions. New preface and appendix. 288pp. 5⅜ x 8½. 0-486-61471-9

ASYMPTOTIC METHODS IN ANALYSIS, N. G. de Bruijn. An inexpensive, comprehensive guide to asymptotic methods–the pioneering work that teaches by explaining worked examples in detail. Index. 224pp. 5⅜ x 8½ 0-486-64221-6

APPLIED COMPLEX VARIABLES, John W. Dettman. Step-by-step coverage of fundamentals of analytic function theory–plus lucid exposition of five important applications: Potential Theory; Ordinary Differential Equations; Fourier Transforms; Laplace Transforms; Asymptotic Expansions. 66 figures. Exercises at chapter ends. 512pp. 5⅜ x 8½. 0-486-64670-X

INTRODUCTION TO LINEAR ALGEBRA AND DIFFERENTIAL EQUATIONS, John W. Dettman. Excellent text covers complex numbers, determinants, orthonormal bases, Laplace transforms, much more. Exercises with solutions. Undergraduate level. 416pp. 5⅜ x 8½. 0-486-65191-6

RIEMANN'S ZETA FUNCTION, H. M. Edwards. Superb, high-level study of landmark 1859 publication entitled "On the Number of Primes Less Than a Given Magnitude" traces developments in mathematical theory that it inspired. xiv+315pp. 5⅜ x 8½. 0-486-41740-9

CATALOG OF DOVER BOOKS

CALCULUS OF VARIATIONS WITH APPLICATIONS, George M. Ewing. Applications-oriented introduction to variational theory develops insight and promotes understanding of specialized books, research papers. Suitable for advanced undergraduate/graduate students as primary, supplementary text. 352pp. 5⅜ x 8½.
0-486-64856-7

COMPLEX VARIABLES, Francis J. Flanigan. Unusual approach, delaying complex algebra till harmonic functions have been analyzed from real variable viewpoint. Includes problems with answers. 364pp. 5⅜ x 8½. 0-486-61388-7

AN INTRODUCTION TO THE CALCULUS OF VARIATIONS, Charles Fox. Graduate-level text covers variations of an integral, isoperimetrical problems, least action, special relativity, approximations, more. References. 279pp. 5⅜ x 8½.
0-486-65499-0

COUNTEREXAMPLES IN ANALYSIS, Bernard R. Gelbaum and John M. H. Olmsted. These counterexamples deal mostly with the part of analysis known as "real variables." The first half covers the real number system, and the second half encompasses higher dimensions. 1962 edition. xxiv+198pp. 5⅜ x 8½. 0-486-42875-3

CATASTROPHE THEORY FOR SCIENTISTS AND ENGINEERS, Robert Gilmore. Advanced-level treatment describes mathematics of theory grounded in the work of Poincaré, R. Thom, other mathematicians. Also important applications to problems in mathematics, physics, chemistry and engineering. 1981 edition. References. 28 tables. 397 black-and-white illustrations. xvii + 666pp. 6⅛ x 9¼.
0-486-67539-4

INTRODUCTION TO DIFFERENCE EQUATIONS, Samuel Goldberg. Exceptionally clear exposition of important discipline with applications to sociology, psychology, economics. Many illustrative examples; over 250 problems. 260pp. 5⅜ x 8½. 0-486-65084-7

NUMERICAL METHODS FOR SCIENTISTS AND ENGINEERS, Richard Hamming. Classic text stresses frequency approach in coverage of algorithms, polynomial approximation, Fourier approximation, exponential approximation, other topics. Revised and enlarged 2nd edition. 721pp. 5⅜ x 8½. 0-486-65241-6

INTRODUCTION TO NUMERICAL ANALYSIS (2nd Edition), F. B. Hildebrand. Classic, fundamental treatment covers computation, approximation, interpolation, numerical differentiation and integration, other topics. 150 new problems. 669pp. 5⅜ x 8½. 0-486-65363-3

THREE PEARLS OF NUMBER THEORY, A. Y. Khinchin. Three compelling puzzles require proof of a basic law governing the world of numbers. Challenges concern van der Waerden's theorem, the Landau-Schnirelmann hypothesis and Mann's theorem, and a solution to Waring's problem. Solutions included. 64pp. 5⅜ x 8½.
0-486-40026-3

THE PHILOSOPHY OF MATHEMATICS: AN INTRODUCTORY ESSAY, Stephan Körner. Surveys the views of Plato, Aristotle, Leibniz & Kant concerning propositions and theories of applied and pure mathematics. Introduction. Two appendices. Index. 198pp. 5⅜ x 8½. 0-486-25048-2

INTRODUCTORY REAL ANALYSIS, A.N. Kolmogorov, S. V. Fomin. Translated by Richard A. Silverman. Self-contained, evenly paced introduction to real and functional analysis. Some 350 problems. 403pp. 5⅜ x 8½.   0-486-61226-0

APPLIED ANALYSIS, Cornelius Lanczos. Classic work on analysis and design of finite processes for approximating solution of analytical problems. Algebraic equations, matrices, harmonic analysis, quadrature methods, much more. 559pp. 5⅜ x 8½.
0-486-65656-X

AN INTRODUCTION TO ALGEBRAIC STRUCTURES, Joseph Landin. Superb self-contained text covers "abstract algebra": sets and numbers, theory of groups, theory of rings, much more. Numerous well-chosen examples, exercises. 247pp. 5⅜ x 8½.   0-486-65940-2

QUALITATIVE THEORY OF DIFFERENTIAL EQUATIONS, V. V. Nemytskii and V.V. Stepanov. Classic graduate-level text by two prominent Soviet mathematicians covers classical differential equations as well as topological dynamics and ergodic theory. Bibliographies. 523pp. 5⅜ x 8½.   0-486-65954-2

THEORY OF MATRICES, Sam Perlis. Outstanding text covering rank, nonsingularity and inverses in connection with the development of canonical matrices under the relation of equivalence, and without the intervention of determinants. Includes exercises. 237pp. 5⅜ x 8½.   0-486-66810-X

INTRODUCTION TO ANALYSIS, Maxwell Rosenlicht. Unusually clear, accessible coverage of set theory, real number system, metric spaces, continuous functions, Riemann integration, multiple integrals, more. Wide range of problems. Undergraduate level. Bibliography. 254pp. 5⅜ x 8½.   0-486-65038-3

MODERN NONLINEAR EQUATIONS, Thomas L. Saaty. Emphasizes practical solution of problems; covers seven types of equations. ". . . a welcome contribution to the existing literature...."–*Math Reviews*. 490pp. 5⅜ x 8½.   0-486-64232-1

MATRICES AND LINEAR ALGEBRA, Hans Schneider and George Phillip Barker. Basic textbook covers theory of matrices and its applications to systems of linear equations and related topics such as determinants, eigenvalues and differential equations. Numerous exercises. 432pp. 5⅜ x 8½.   0-486-66014-1

LINEAR ALGEBRA, Georgi E. Shilov. Determinants, linear spaces, matrix algebras, similar topics. For advanced undergraduates, graduates. Silverman translation. 387pp. 5⅜ x 8½.   0-486-63518-X

ELEMENTS OF REAL ANALYSIS, David A. Sprecher. Classic text covers fundamental concepts, real number system, point sets, functions of a real variable, Fourier series, much more. Over 500 exercises. 352pp. 5⅜ x 8½.   0-486-65385-4

SET THEORY AND LOGIC, Robert R. Stoll. Lucid introduction to unified theory of mathematical concepts. Set theory and logic seen as tools for conceptual understanding of real number system. 496pp. 5⅜ x 8¼.   0-486-63829-4

TENSOR CALCULUS, J.L. Synge and A. Schild. Widely used introductory text covers spaces and tensors, basic operations in Riemannian space, non-Riemannian spaces, etc. 324pp. 5⅜ x 8¼. 0-486-63612-7

ORDINARY DIFFERENTIAL EQUATIONS, Morris Tenenbaum and Harry Pollard. Exhaustive survey of ordinary differential equations for undergraduates in mathematics, engineering, science. Thorough analysis of theorems. Diagrams. Bibliography. Index. 818pp. 5⅜ x 8½. 0-486-64940-7

INTEGRAL EQUATIONS, F. G. Tricomi. Authoritative, well-written treatment of extremely useful mathematical tool with wide applications. Volterra Equations, Fredholm Equations, much more. Advanced undergraduate to graduate level. Exercises. Bibliography. 238pp. 5⅜ x 8½. 0-486-64828-1

FOURIER SERIES, Georgi P. Tolstov. Translated by Richard A. Silverman. A valuable addition to the literature on the subject, moving clearly from subject to subject and theorem to theorem. 107 problems, answers. 336pp. 5⅜ x 8½. 0-486-63317-9

INTRODUCTION TO MATHEMATICAL THINKING, Friedrich Waismann. Examinations of arithmetic, geometry, and theory of integers; rational and natural numbers; complete induction; limit and point of accumulation; remarkable curves; complex and hypercomplex numbers, more. 1959 ed. 27 figures. xii+260pp. 5⅜ x 8½. 0-486-63317-9

POPULAR LECTURES ON MATHEMATICAL LOGIC, Hao Wang. Noted logician's lucid treatment of historical developments, set theory, model theory, recursion theory and constructivism, proof theory, more. 3 appendixes. Bibliography. 1981 edition. ix + 283pp. 5⅜ x 8½. 0-486-67632-3

CALCULUS OF VARIATIONS, Robert Weinstock. Basic introduction covering isoperimetric problems, theory of elasticity, quantum mechanics, electrostatics, etc. Exercises throughout. 326pp. 5⅜ x 8½. 0-486-63069-2

THE CONTINUUM: A CRITICAL EXAMINATION OF THE FOUNDATION OF ANALYSIS, Hermann Weyl. Classic of 20th-century foundational research deals with the conceptual problem posed by the continuum. 156pp. 5⅜ x 8½. 0-486-67982-9

CHALLENGING MATHEMATICAL PROBLEMS WITH ELEMENTARY SOLUTIONS, A. M. Yaglom and I. M. Yaglom. Over 170 challenging problems on probability theory, combinatorial analysis, points and lines, topology, convex polygons, many other topics. Solutions. Total of 445pp. 5⅜ x 8½. Two-vol. set.
Vol. I: 0-486-65536-9   Vol. II: 0-486-65537-7

INTRODUCTION TO PARTIAL DIFFERENTIAL EQUATIONS WITH APPLICATIONS, E. C. Zachmanoglou and Dale W. Thoe. Essentials of partial differential equations applied to common problems in engineering and the physical sciences. Problems and answers. 416pp. 5⅜ x 8½. 0-486-65251-3

THE THEORY OF GROUPS, Hans J. Zassenhaus. Well-written graduate-level text acquaints reader with group-theoretic methods and demonstrates their usefulness in mathematics. Axioms, the calculus of complexes, homomorphic mapping, *p*-group theory, more. 276pp. 5⅜ x 8½. 0-486-40922-8

HYDRODYNAMIC AND HYDROMAGNETIC STABILITY, S. Chandrasekhar. Lucid examination of the Rayleigh-Benard problem; clear coverage of the theory of instabilities causing convection. 704pp. 5⅜ x 8¼.   0-486-64071-X

INVESTIGATIONS ON THE THEORY OF THE BROWNIAN MOVEMENT, Albert Einstein. Five papers (1905–8) investigating dynamics of Brownian motion and evolving elementary theory. Notes by R. Fürth. 122pp. 5⅜ x 8½.   0-486-60304-0

THE PHYSICS OF WAVES, William C. Elmore and Mark A. Heald. Unique overview of classical wave theory. Acoustics, optics, electromagnetic radiation, more. Ideal as classroom text or for self-study. Problems. 477pp. 5⅜ x 8½.   0-486-64926-1

GRAVITY, George Gamow. Distinguished physicist and teacher takes reader-friendly look at three scientists whose work unlocked many of the mysteries behind the laws of physics: Galileo, Newton, and Einstein. Most of the book focuses on Newton's ideas, with a concluding chapter on post-Einsteinian speculations concerning the relationship between gravity and other physical phenomena. 160pp. 5⅜ x 8½.
0-486-42563-0

PHYSICAL PRINCIPLES OF THE QUANTUM THEORY, Werner Heisenberg. Nobel Laureate discusses quantum theory, uncertainty, wave mechanics, work of Dirac, Schroedinger, Compton, Wilson, Einstein, etc. 184pp. 5⅜ x 8½.0-486-60113-7

ATOMIC SPECTRA AND ATOMIC STRUCTURE, Gerhard Herzberg. One of best introductions; especially for specialist in other fields. Treatment is physical rather than mathematical. 80 illustrations. 257pp. 5⅜ x 8½.   0-486-60115-3

AN INTRODUCTION TO STATISTICAL THERMODYNAMICS, Terrell L. Hill. Excellent basic text offers wide-ranging coverage of quantum statistical mechanics, systems of interacting molecules, quantum statistics, more. 523pp. 5⅜ x 8½.
0-486-65242-4

THEORETICAL PHYSICS, Georg Joos, with Ira M. Freeman. Classic overview covers essential math, mechanics, electromagnetic theory, thermodynamics, quantum mechanics, nuclear physics, other topics. First paperback edition. xxiii + 885pp. 5⅜ x 8½.   0-486-65227-0

PROBLEMS AND SOLUTIONS IN QUANTUM CHEMISTRY AND PHYSICS, Charles S. Johnson, Jr. and Lee G. Pedersen. Unusually varied problems, detailed solutions in coverage of quantum mechanics, wave mechanics, angular momentum, molecular spectroscopy, more. 280 problems plus 139 supplementary exercises. 430pp. 6½ x 9¼.   0-486-65236-X

THEORETICAL SOLID STATE PHYSICS, Vol. 1: Perfect Lattices in Equilibrium; Vol. II: Non-Equilibrium and Disorder, William Jones and Norman H. March. Monumental reference work covers fundamental theory of equilibrium properties of perfect crystalline solids, non-equilibrium properties, defects and disordered systems. Appendices. Problems. Preface. Diagrams. Index. Bibliography. Total of 1,301pp. 5⅜ x 8½. Two volumes.   Vol. I: 0-486-65015-4   Vol. II: 0-486-65016-2

WHAT IS RELATIVITY? L. D. Landau and G. B. Rumer. Written by a Nobel Prize physicist and his distinguished colleague, this compelling book explains the special theory of relativity to readers with no scientific background, using such familiar objects as trains, rulers, and clocks. 1960 ed. vi+72pp. 5⅜ x 8½.   0-486-42806-0

CATALOG OF DOVER BOOKS

A TREATISE ON ELECTRICITY AND MAGNETISM, James Clerk Maxwell. Important foundation work of modern physics. Brings to final form Maxwell's theory of electromagnetism and rigorously derives his general equations of field theory. 1,084pp. 5⅜ x 8½. Two-vol. set. Vol. I: 0-486-60636-8 Vol. II: 0-486-60637-6

QUANTUM MECHANICS: PRINCIPLES AND FORMALISM, Roy McWeeny. Graduate student-oriented volume develops subject as fundamental discipline, opening with review of origins of Schrödinger's equations and vector spaces. Focusing on main principles of quantum mechanics and their immediate consequences, it concludes with final generalizations covering alternative "languages" or representations. 1972 ed. 15 figures. xi+155pp. 5⅜ x 8½. 0-486-42829-X

INTRODUCTION TO QUANTUM MECHANICS With Applications to Chemistry, Linus Pauling & E. Bright Wilson, Jr. Classic undergraduate text by Nobel Prize winner applies quantum mechanics to chemical and physical problems. Numerous tables and figures enhance the text. Chapter bibliographies. Appendices. Index. 468pp. 5⅜ x 8½. 0-486-64871-0

METHODS OF THERMODYNAMICS, Howard Reiss. Outstanding text focuses on physical technique of thermodynamics, typical problem areas of understanding, and significance and use of thermodynamic potential. 1965 edition. 238pp. 5⅜ x 8½. 0-486-69445-3

THE ELECTROMAGNETIC FIELD, Albert Shadowitz. Comprehensive undergraduate text covers basics of electric and magnetic fields, builds up to electromagnetic theory. Also related topics, including relativity. Over 900 problems. 768pp. 5⅝ x 8¼. 0-486-65660-8

GREAT EXPERIMENTS IN PHYSICS: FIRSTHAND ACCOUNTS FROM GALILEO TO EINSTEIN, Morris H. Shamos (ed.). 25 crucial discoveries: Newton's laws of motion, Chadwick's study of the neutron, Hertz on electromagnetic waves, more. Original accounts clearly annotated. 370pp. 5⅜ x 8½. 0-486-25346-5

EINSTEIN'S LEGACY, Julian Schwinger. A Nobel Laureate relates fascinating story of Einstein and development of relativity theory in well-illustrated, nontechnical volume. Subjects include meaning of time, paradoxes of space travel, gravity and its effect on light, non-Euclidean geometry and curving of space-time, impact of radio astronomy and space-age discoveries, and more. 189 b/w illustrations. xiv+250pp. 8⅜ x 9¼. 0-486-41974-6

STATISTICAL PHYSICS, Gregory H. Wannier. Classic text combines thermodynamics, statistical mechanics and kinetic theory in one unified presentation of thermal physics. Problems with solutions. Bibliography. 532pp. 5⅜ x 8½. 0-486-65401-X